TEACHER'S ANNOTA'

Exploring Circles

An Alternative Unit for Analyzing Circles

Author
David Foster

GLENCOE
McGraw-Hill

New York, New York
Columbus, Ohio
Mission Hills, California
Peoria, Illinois

Copyright © 1996 by Glencoe/McGraw-Hill. All rights reserved.
Except as permitted under the United States Copyright Act, no part of this publication may be reproduced or distributed in any form or by any means, or stored in a database or retrieval system, without prior written permission of the publisher.

Printed in the United States of America.

Send all inquiries to:
Glencoe/McGraw-Hill
936 Eastwind Drive
Westerville, OH 43081

ISBN: 0-02-824218-1 (Student Edition)
ISBN: 0-02-824220-3 (Teacher's Annotated Edition)

1 2 3 4 5 6 7 8 9 10 VH/LH-P 04 03 02 01 00 99 98 97 96 95

TEACHER'S ANNOTATED EDITION

Table of Contents

Teacher's Handbook — **T4**
- Using the Unit — T4
- Instructional Model — T4
- Grouping Your Students — T5
- Assessing Your Students — T5
- Using Journals and Portfolios — T6
- Student Diversity — T7
- Key to Icons in Margin Column — T7

Unit Overview — **T8**
- Why This Unit Is Important — T8
- Investigation Profile — T8
- Unit Outcomes — T9

Using *Exploring Circles* with *Merrill Pre-Algebra*, *Merrill Algebra 1*, *Merrill Geometry*, and *Merrill Algebra 2* — **T9**

Using Manipulatives and Materials — **T10**

Teaching Suggestions and Strategies — **T11**
- Investigation 1 — T11
- Investigation 2 — T15
- Investigation 3 — T18
- Investigation 4 — T21
- Investigation 5 — T24

Teacher's Answer Key — **T27**

Teacher's Handbook

Using the Unit

This Teacher's Annotated Edition includes the Student Edition together with Teacher Notes and suggested solutions. The unit consists of five investigations, which are each made up of activities. Each activity is designed around the following instructional model, which is an adaptation and extension of one used in the Middle Grades Mathematics Project (MGMP). The MGMP is a program developed at Michigan State University to develop mathematics units for instruction in grades 5 through 8. The National Science Foundation funded the program. This model, like the unit itself, requires new roles for both students and teachers.

Instructional Model

Motivate and Focus	During this first step, the teacher initiates whole-class discussion of the problem situation or question(s) posed at the beginning of each investigation and related activity. This sets the context for the student work to follow. It provides an opportunity to clarify directions for the group activities. The teacher is the director and moderator.
Explore	The next step involves focused problems/questions related to the *Motivate and Focus* situation where investigation leads to gathering data, looking for patterns, and making conjectures. Students work cooperatively in small groups. While groups are working, the teacher circulates from group to group providing guidance and support. This may entail clarifying or asking questions, giving hints, providing encouragement, and drawing group members into the discussion to help groups work more cooperatively. The materials drive the learning, and the teacher is the facilitator.
Share and Summarize	This utilizes whole-class discussion of results found by different small groups. It leads to a whole-class summary of important ideas or to further exploration of a topic if competing perspectives remain. Varying points of view and differing conclusions that can be justified should be encouraged. The teacher is the moderator.
Assess	A group task is used to reinforce initial understanding of the concept or method. The teacher circulates around the room assessing levels of understanding. The teacher is the intellectual coach. Depending on the goals of the activity, there may be additional *Explore, Share and Summarize, Assess* sequences. An overall class discussion and synthesis by the teacher should occur at the end of each investigation.
Related Applications	This involves a set of related or new contexts to which student-developed ideas and methods can be applied. A subset of these additional applications should be assigned for further group work in class and/or for individual work outside of class. Decisions should be based on student performance and the availability of time and technology. Students should exercise some choice of applications to pursue; at times they should be given the opportunity to pose their own problems/questions to investigate. The teacher circulates around the room acting as facilitator and coach.
Extensions	Finally, students work on exercises that permit further or deeper/more formal study of the topic under investigation. Every student should have the opportunity to complete one or more of these extensions during the course of a unit.

● Grouping Your Students

This unit has been designed for use in classrooms containing students of diverse backgrounds. In order to ensure heterogeneous grouping, you should make the final decision about group assignments based on mathematical performance levels, gender, and cultural backgrounds at the beginning of each investigation. Students can be re-grouped later in an investigation for related applications and extensions.

If your students have not previously experienced small-group cooperative learning, you will need to present some guidelines for group behavior. These should include:

- Listen carefully and with respect to each other and try, whenever possible, to build on the ideas of each other.
- Make sure everyone contributes to the group task and no one person dominates.
- Ask for clarification or help when something is not understood and help others in the group when asked.
- Achieve a group answer or solution for each task.
- Make sure everyone understands the solution before sharing it with the class or before the group goes on to the next task.

Expect that group work will go more smoothly on some days than others. It usually takes about three weeks for students to begin functioning well in groups. From time to time, remind students of the above guidelines for cooperative group behavior.

More information about cooperative learning can be found in Glencoe's *Cooperative Learning in the Mathematics Classroom*.

● Assessing Your Students

Instruction and assessment are closely linked. Both the Student Edition and Teacher's Annotated Edition provide tasks for group as well as individual assessment. The Teaching Suggestions and Strategies also provide suggestions for methods of informal assessment of student performance based on observations. The nature of the Student Edition and the small group format are ideally suited to assessment of mathematical thinking, communication, and disposition toward mathematics through student observations.

For more information on performance assessment, see Glencoe's *Alternative Assessment in the Mathematics Classroom*.

● Using Journals and Portfolios

Journals and portfolios are important forms of assessment. Students will find journal prompts and suggestions for what to include in a portfolio in the margins of the activities.

Journals A journal is a written account that a student keeps to record what she or he has learned. Journal entries are conducive to thinking about why something has been done. They can be used to record and summarize key topics studied, the student's feelings toward mathematics, accomplishments or frustrations in solving a particular problem or topic, or any other notes or comments the student wishes to make. Keeping a mathematical journal can be helpful in students' development of a reflective and introspective point of view. It also encourages students to have a more thoughtful attitude toward written work and should be instrumental in helping students learn more mathematics. Journals are also an excellent way for students to practice and improve their writing skills.

Portfolios A portfolio is a representative sample of a student's work that is collected over a period of time. The selection of work samples for a portfolio should be done with an eye toward presenting a balanced portrait of a student's achievements. The pieces of work placed in a portfolio should have more significance than other work a student has done. They are chosen as illustrations of a student's best work at a particular point in time. Thus, the range of items selected shows a student's intellectual growth in mathematics over time.

Students may select the products of activities for inclusion. Bear in mind that the actual selection of the items by the students will tell you what pieces of work the students think are significant. In addition, students should reflect upon their selections by explaining why each particular work was chosen.

The following examples illustrate topics that would be appropriate for inclusion in a portfolio.

- a solution to a difficult or nonroutine problem that shows originality of thought
- a written report of an individual project or investigation
- examples of problems or conjectures formulated by the student
- mathematical artwork, charts, or graphs
- a student's contribution to a group report or investigation
- a photo or sketch of physical models or manipulative
- statements on mathematical disposition, such as motivation, curiosity, and self confidence
- a first and final draft of a piece of work that shows student growth

Student Diversity

In the *Exploring Circles* activities, students are asked to interpret data to discuss and support approaches to problems. Students will have to create tables, write reports, defend solutions, and draw conclusions. The students in your class may not have the same level of proficiency required to carry out and produce such high level forms of communication. Reasons for this diversity include lack of proficiency in English, poor knowledge of mathematical terminology, limited exposure to the rules and use of language in mathematical contexts, and lack of background and experience with technical forms of communication.

To foster participation and effective communication by *all* students, we must try to obtain an idea of the students' communication skills. This can be accomplished by using the students' responses to each investigation opener. Listen to the students as they talk about and interpret the task. Who is having difficulties understanding the requirements? Which students cannot identify and explain the components and objectives of the activity?

Second, examine the written work. Which students cannot fully explain the solution? Is this due to a lack of mathematical knowledge or the inability to express thoughts in written or oral form?

Lastly, you may wish to use the following general strategies to address communication diversity issues.

- Spend the initial phase of the activities clarifying terms, expressions, task requirements, and symbols. Encourage students to offer their interpretations of the problems or task components.

- Encourage students to describe ideas and processes orally in their cooperative learning groups. Many communication skills are developed through practice.

- Let students use their first language in small groups if they are unable to fully communicate mathematical ideas in English.

Key to Icons in Margin Column

FYI	**FYIs** are "fast facts," entertaining tidbits, and fascinating math-related trivia.
Graphing Calculator Activity	*Exploring Circles* contains 4 **Graphing Calculator Activities** that allow students to explore mathematical concepts.
Share and Summarize	These headings suggest class discussion about the results found by different groups.
Portfolio Assessment	A **Portfolio Assessment** suggestion asks students to select items from their work that represent their knowledge.
Journal	A **Journal Entry** appears in many of the activities, giving students the opportunity to keep a written log of their thoughts.

Exploring Circles

Unit Overview

● **Why This Unit is Important**

The circle may be the most basic and most complex figure in geometry. Students in their early years learn basic properties of the circle, yet they continue to revisit circles throughout the study of calculus and beyond. The properties of circles are fundamental to geometry and are illustrated in the following quote from *On the Shoulders of Giants: New Approaches to Numeracy*.

> "The fundamental geometry problem for one-dimensional phenomena is the determination of distance along a path. Key examples include calculation or comparison of perimeters of curves and polygons. There is one geometric number—π—that all students should learn to understand."

Throughout this unit, students will experience mathematics that is relevant to their lives. Understanding the properties of geometry that involve circles and circular properties are fundamental to being mathematically powerful. Students will explore important mathematics and reach a deeper understanding of the properties of circles.

● **Investigation Profile**

Investigation 1 Students investigate the perimeters of regular polygons and make a conjecture regarding the circumference of circles. Students develop methods of measuring circumference and display their data on graphs. The relationship between the diameter of a circle and its circumference is explored. Students investigate circles and functions.

Investigation 2 Students develop a formula for the area of a circle by examining the areas of regular polygons. Students use triangles in regular polygons to generalize a formula for finding the area of a circle. Students investigate the relationship between areas of polygons and their perimeters and develop a conjecture about maximizing area with a fixed perimeter. Students also investigate how to maximize the number of circles that can be cut from a quadrilateral.

Investigation 3 Students use historical methods to develop approximations of π. Technology is used to calculate π using coordinate geometry. The students also investigate infinite series and products to calculate approximations of π. Students compare various methods for calculating the approximation of π and determine which methods are most accurate and efficient.

Investigation 4 This investigation allows students to explore parts of a circle and concentric circles. The students use radians and degrees to find the length of arcs and the areas of sectors. They compare the areas of sectors and circles. Students use problems taken from real life to explore the relationships among arcs, sectors, and concentric circles.

Investigation 5 This investigation involves the use of computers and the LOGO programming language. Using LOGO, students explore circles and inscribed angles. They develop the relationship between central angles and inscribed angles in a circle. Students also use straightedge, compass, and protractor to explore the relationships of angles in circles. LOGO is used to construct regular polygons and stars and to develop a method for determining their interior and exterior angles.

Unit Outcomes

There are four fundamental outcomes of the work in *Exploring Circles*.

- Students will understand properties of circles and their relationship with regular polygons.
- Students will investigate procedures for approximating π.
- Students will use circles and their properties as tools in solving problems.
- Students will calculate important measures of angles inscribed in circles and understand the use of those measures in solving problems.

Using *Exploring Circles* with
Merrill Pre-Algebra: A Transition to Algebra
Merrill Algebra 1: Applications and Connections,
Merrill Geometry: Applications and Connections, and
Merrill Algebra 2 With Trigonometry: Applications and Connections

Exploring Circles consists of five investigations that may be used as alternative or supplemental materials for *Merrill Pre-Algebra, Merrill Algebra 1, Merrill Geometry* and *Merrill Algebra 2*. The following correlation shows you which investigation can be used with lessons in each text.

Exploring Circles Investigation Number	Merrill Pre-Algebra Lesson Number	Merrill Algebra 1 Lesson Number	Merrill Geometry Lesson Number	Merrill Algebra 2 Lesson Number
1	3-6, 6-3	1-7	9-1, 10-6, 10-7	page 733
2	13-1, 13-3, 13-5	1-1, 1-7	10-5, 10-6, 10-7	12-6
3	page 245		page 430	12-6
4		page 479	9-2	16-1
5			page 49, 9-3, 9-4	page 227

Using Manipulatives and Materials

Cooperative learning activities, which often include the use of manipulatives, are the basis for all learning in *Exploring Circles*. Below is a list of the manipulatives and materials suggested for this unit. A list specific to each activity also appears at the beginning of the activity.

Manipulative	Activity	Manipulative	Activity
Centimeter Graph paper	1-1, 1-4, 2-1, 2-2, 2-3, 3-2	Graphing Calculator	1-1, 3-3
Centimeter Ruler	1-1, 1-4, 4-1	Tape Measure	1-2, 1-3
Cylindrical Objects	1-2, 1-3	Scissors	1-3, 4-1
String	1-3, 3-1	Colored Pencils	1-4
Eyedropper	1-4	Paper Towels	1-4
Tiles	2-2	Calculator	2-2, 4-2, 4-3
Pennies	2-3	Poster Paper	3-1
Ribbon	3-1	Meter Stick	3-1
Compass	3-2, 5-3	Blank Transparency	3-2
Bobby Pins	3-2	Pipe Cleaners	4-1
Logo Software	5-1, 5-2, 5-4	Map	5-1
Straightedge	5-3	Protractor	5-3

Teaching Suggestions and Strategies

Exploring Circumference

Mathematical Overview

The activities include problems related to real-life situations involving circles. Students investigate the perimeters of regular polygons and make a conjecture about the circumference of circles. Students develop methods of measuring circumference and display their data on graphs. They explore the relationship between the diameter of a circle and its circumference. Students investigate circles and functions. They use graphing calculators to investigate conjectures.

Investigation Outcomes

Students will
- explore perimeters of regular figures;
- describe the relationship between the circumference and the diameter of circles;
- use graphs and functions to investigate properties of circles.

Materials Needed

graph paper

cylindrical objects

string

paper towels

graphing calculator

Blackline Masters 1-1A, 1-1B

centimeter ruler

tape measure

eye dropper

colored pencils

scissors

Teaching Suggestions and Strategies **T11**

Activity 1-1   Perimeters of Regular Polygons

1 Motivate and Focus

Ask students to name some real objects in the shape of regular polygons. **Sample answers: road signs, pieces of leather on a soccer ball, and tiles**

2 Share and Summarize

Have students read and discuss the opening problem involving designing a new arena at a university to optimize the number of good seats. Discuss with the class some attributes of a good arena. Have students share their experiences. Focus the discussion on the task of finding the arena design that will allow for the best seating.

In Exercise 1a students are asked to find the **perpendicular bisector** of two adjacent sides of each polygon on Blackline Master 1-1A. A perpendicular bisector is a line or line segment that passes through the midpoint of a segment and is perpendicular to that segment. Students can find the perpendicular bisector of a side of a polygon by first cutting out the polygon. Then have them fold the polygon so that one vertex falls on an adjacent vertex as shown below. This fold is the perpendicular bisector.

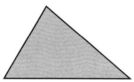 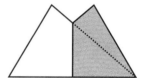

Tell students to unfold the polygon and repeat the process with the vertices of an adjoining side. The intersection of the folds is the center of the poygon.

Bring the students together after Exercises 1e and 3c to share their data and discuss their findings. Use the data from the tables to make conjectures.

3 Extend the Activity

Students may create three-dimensional designs and create models of the arena.

4 Assess the Activity

Assessment should be ongoing to judge direction and pace of the instruction and to clarify the students' thinking and evaluate their mathematical growth.

Regularly question students to assess their understanding and to help them clarify their thinking. This should be the environment of the class. Students should question each other, and the teacher should question students individually, within small groups as well as with the entire class. Modification of the direction and pace of the curriculum should follow the assessment.

For this activity, students should make a written recommendation to the designers of the arena. The work should be assessed on the mathematics the student used in the design and the completeness of the justification of the design.

Activity 1-2 The Relationship of Circumference and Diameter

1 Motivate and Focus

Ask students what they know or remember about π. Have students press $\boxed{\pi}$ on their calculators. Tell them that in the fifth century, a Chinese mathematician developed the rational approximation $\frac{355}{113}$ for π. Have students divide 355 by 113 and compare the quotient with the calculated approximation.

2 Share and Summarize

Have students read and discuss the problem involving designing a circular swimming pool to optimize the distance across the pool. Discuss with the class key ideas in the problem. Have students share their experiences. Have the students focus on the task of finding the relationship between the diameter and circumference of a circle.

Bring the students together after Exercises 2d and 3f to share their data and discuss their findings. Invite students to share their understanding of the problem.

3 Extend the Activity

Students may investigate the reason why pools are designed in the shapes they are and list the advantages and disadvantages of differently-shaped pools.

4 Assess the Activity

Use Exercise 3f as an assessment. The work should be assessed on the mathematics the student used in the design and the completeness of the justification of the design.

Activity 1-3 More on Circumference and Diameter Relationship

1 Motivate and Focus

Have students explore differently-sized circular objects and discuss their discoveries. Having hands-on experiences gives students insight into solving the problem.

2 Share and Summarize

Have students read and discuss the problem involving quality control for a barrel manufacturing company. Discuss with the class key ideas in the problem.

Bring students together after Exercise 2b. Invite students to share their understanding of the problem.

Teaching Suggestions and Strategies

3 Extend the Activity

Students could explore the graphs (Exercise 4) representing their data by utilizing technology.

4 Assess the Activity

Use Exercise 3 as an assessment. The work should be assessed on the mathematics the student uses in the report and the completeness of the justification of the report.

Activity 1-4   Circles and Functions

1 Motivate and Focus

Ask students to consider the statement "Your hat size is a function of the circumference of your head." Develop the idea that "is a function of" actually means "depends on." Have students give examples of additional situations in which the value of one variable results in only one true value for a second variable.

2 Share and Summarize

Bring students together after Exercise 1g and again after Exercise 3. Invite them to share their understanding of the problem and how the experiment relates to it.

3 Extend the Activity

After students research the method used to find the rainfall for a region, ask them to design an experiment that could be used to measure the amount of simulated rain from a lawn sprinkler.

4 Assess the Investigation

Have students design a small experiment in which they choose two variables they think might be related. Have them collect the data on those variables, construct tables and graphs, and decide whether the relationship between those two variables represents a function. Ask for completeness and careful student descriptions of all aspects of the experiment.

T14   Exploring Circles

Exploring Area

● **Mathematical Overview**

In this investigation, students use triangles to generalize a formula for finding the area of a regular polygon. Students investigate the relationship between area and perimeter of a rectangle while the area stays the same. Then they work to maximize the area of a rectangle while experimenting with its dimensions. In the last activity, students explore ways to maximize the number of circles that can be cut from a quadrilateral.

● **Investigation Outcomes**

Students will

- explore areas of regular polygons;
- describe the relationship between the area and perimeter of rectangles;
- use mathematical techniques to maximize area of polygons.

● **Materials Needed**

centimeter grid paper calculator

tiles pennies

Blackline Master 2-1 Transparency Master 2-2

Activity 2-1 Area of Regular Polygons

1 Motivate and Focus

Ask students to state any formulas they know and what the formulas are used for. List these on the chalkboard, identifying the variables. Have them evaluate each formula after choosing values for the variables.

2 Share and Summarize

Have students read and discuss the problem to investigate if there is a single method for finding the area of a regular polygon. Discuss with the class key ideas in the problem. Discuss the meaning of area. Bring students together after Exercise 4d.

3 Extend the Activity

Have students research the dimensions of a standard stop sign. Then have them draw and label the dimensions of the three "new" signs if the area of each sign is equal to the area of the stop sign.

Teaching Suggestions and Strategies **T15**

4 Assess the Activity

Assessment should be ongoing to judge direction and pace of the instruction and to clarify the students' thinking and evaluate their mathematical growth.

Regularly question students to assess their understanding and to help them clarify their thinking. This should be the environment of the class. Students should question each other, and the teacher should question students individually, within small groups as well as with the entire class. Modification of the direction and pace of the curriculum should follow the assessment.

Have students find the area of each of the following polygons.
- square with a side of 4.5 cm and an apothem of 2.25 cm   **20.25 cm²**
- regular pentagon with a side of 6 meters and an apothem of 4.13 meters   **61.95 m²**
- regular octagon with a side of 5 inches and an apothem of 6.035 inches   **120.7 in²**
- regular dodecagon (12 sides) with a side of 12 feet and an apothem of 18.5 feet   **1,332 ft²**

Activity 2-2   Maximizing Area

1 Motivate and Focus

Remind students that perimeter is the distance around a figure and area is the measure of the surface enclosed by a figure. Ask them to share ways they remember the difference.

2 Share and Summarize

After students complete Exercise 1, bring them together to discuss their findings. Work out any disagreements and gain a consensus. If students are having difficulty with Exercise 2, show them Transparency Master 2-2.

Share and summarize again after Exercises 3 and 4.

3 Extend the Activity

This activity could be done as a spreadsheet application. This would allow students to immediately see the impact of change in dimensions upon area. It also allows them to play "What if?" with their data.

4 Assess the Activity

Exercise 3d can be used as an assessment. The reports should be assessed on the mathematics used in the report and the completeness of the report.

Activity 2-3 Circles and Quadrilaterals

1 Motivate and Focus

Display the illustration at the right and ask students to explain an easy way to find the area of the shaded region.

2 Share and Summarize

Following group discussion of the first paragraph, have students share ideas to ensure group understanding.

Bring students together after Exercise 1d to discuss their reasoning and to develop a class consensus.

3 Extend the Activity

Ask students to write a paragraph describing what they might do with the scraps of metal that are left after cutting out the tops and how it might be useful for a can manufacturing company.

4 Assess the Investigation

Have students find the area of each shaded region below. All the polygons are regular.

a.

78.27 units²

b.

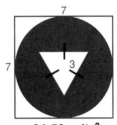

34.59 units²

c.

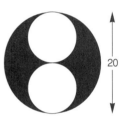

157.08 units²

Teaching Suggestions and Strategies

Exploring π

● **Mathematical Overview**

The activities in this investigation emphasize the historical procedures that have been used to approximate π. The entire investigation is an excellent opportunity for using graphing technology in the instructional process.

● **Investigation Outcomes**

Students will
- use and compare various procedures for approximating π;
- determine which procedure for approximating π is most accurate and most efficient.

● **Materials Needed**

 poster paper centimeter grid paper

 string meter stick

 graphing calculator ribbon

 blank transparency bobby pins

 Transparency Master 3-2 compass

 Blackline Masters 3-2A, 3-2B

Activity 3-1   Inscribing Regular Polygons

1 Motivate and Focus

Ask students what an inscription is, or what it means to inscribe something. **Sample answer: writing on something, like a piece of jewelry.**

Relate this to an inscribed polygon, which is "a polygon written on a circle," or a polygon whose vertices lie on the circle.

2 Share and Summarize

If students have difficulty making their homemade tape measures, tell them to repeatedly fold the unit diameter in half and mark each fold.

Bring students together after Exercise 4c. Discuss any differences in their estimates for the perimeter of the 32-sided polygon. If any groups used the graph or the table for their estimate, ask them to explain how they determined the estimate.

T18   Exploring Circles

Have students share and summarize again after Exercise 4g. Discuss the advantages Archimedes may have had while making his inscribed polygons. **Sample answer: Since his circle was larger, the sides of the 32-sided polygon would have been easier to measure.**

3 Extend the Activity

Have students research and report on other discoveries for which Archimedes is responsible.

Students might also be encouraged to write a report on the approximation techniques of other individuals.

4 Assess the Activity

Assessment should be ongoing to judge direction and pace of the instruction and to clarify students' thinking and evaluate their mathematical growth.

Regularly question students to assess their understanding and to help them clarify their thinking. This should be the environment of the class. Students should question each other, and the teacher should question students individually, within small groups, as well as with the entire class. Modification of the direction and pace of the curriculum should follow the assessment.

For this activity, students write a report about their experiences while reenacting Archimedes' method for approximating π. The reports should be assessed based on the mathematics the students used in their designs and the completeness of their constructions.

Activity 3-2   Using Coordinate Geometry to Calculate π

1 Motivate and Focus

Ask students to describe how they could theoretically figure out the probability of a dart landing on any region of a dartboard. Ask them to discuss how probability can be used to design dartboards and develop a scoring system.

2 Share and Summarize

Ask students to respond to Exercise 3a as a whole-class activity. Students should conclude that the ratio will only *approach* π because it is physically impossible for a dart to land on every point on the dartboard.

Before students remove the transparency in Exercise 6e, have them draw the *x*- and *y*-axes on it. Then display Transparency 3-2 and ask them to superimpose their "dartboards" on it. Discuss the implications of doing this.

Bring students together to share and summarize after Exercise 6f and again after Exercise 7e.

Teaching Suggestions and Strategies

3 Extend the Activity

Ask students to use calculators to see who can find the decimal equivalent of a fraction that comes closest to the value displayed when they press $\boxed{\pi}$ on their calculator.

4 Assess the Activity

Exercise 7e can be used as an assessment. Each report should be assessed on the mathematics used in the report and the completeness of the report.

Activity 3-3   More Methods that Generate π

1 Motivate and Focus

Have students press $\boxed{\pi}$ on their calculators. Tell students to refer to this decimal as they compare the quotients of the rational approximations for π in Exercise 1.

2 Share and Summarize

Whole class discussion should be held after Exercise 2d. Discuss student opinions on what constitutes an efficient or accurate method for approximating π.

3 Extend the Activity

Have students develop their own infinite series or infinite product that approximates π.

4 Assess the Investigation

Ask students to write a paragraph describing what they learned in this investigation and how it might be useful outside of class.

Exploring Arcs and Sectors

● **Mathematical Overview**

This investigation introduces students to arcs, radians, sectors, and concentric circles. Students explore their relationships as they solve real-world problems.

● **Investigation Outcomes**

Students will

- change angle measures from radians to degrees, and vice versa;
- calculate the area of a circle;
- examine the differences between the areas of concentric circles.

● **Materials Needed**

 pipe cleaners

 scissors

 centimeter ruler

 calculator

Activity 4-1   Arcs and Radians

1 Motivate and Focus

Tell students that a plane is flying west. An air traffic controller tells the pilot to turn the plane so that it has a heading of 125° west of south. Ask students to draw a diagram of this situation and determine what angle the airplane turns.   **The plane should turn 35° to the right.**

2 Share and Summarize

To assist students in making their wheels, you may also want to provide a metal washer to act as the "axle" or center of each wheel. Bring students together after the hands-on experience.

After completing Exercise 5, discuss with the class how to change degrees to radians and radians to degrees. Invite students to share their methods for determining a formula or process for converting between radians and degrees.

3 Extend the Activity

Tell students that mathematicians say that an arc *subtends* its central angle. Then tell them that the length of any circular arc is the product of the measure of the radius of the circle and the radian measure of the central angle. Ask students to find the length of an arc that subtends a central angle of 36° in a circle of radius 8 cm.   **The length of the arc is approximately 5.03 centimeters.**

Teaching Suggestions and Strategies   **T21**

4 Assess the Activity

Assessment should be ongoing to judge direction and pace of the instruction and to clarify students' thinking and evaluate their mathematical growth.

Regularly question students to assess their understanding and to help them clarify their thinking. This should be the environment of the class. Students should question each other, and the teacher should question students individually, within small groups, as well as with the entire class. Modification of the direction and pace of the curriculum should follow the assessment.

Students should be able to develop a simple method for converting radians to degrees and vice versa.

Activity 4-2   The Area of Circles and Sectors

1 Motivate and Focus

Have students read the first paragraph of the activity. Then share the following story with the class. *Davina had some coupons. She could get one jumbo pizza for $8.99 or two large pizzas for $9.99. She wants to know which offer is the better deal.* Ask students what Davina needs to find in order to solve the problem. **She needs to find the area of each size of pizza.**

2 Share and Summarize

Ask students to respond to Exercise 5a as a whole-class activity. The profit for selling a pizza by the slice depends on which price is used for the whole pizza price. Students should come to a consensus as to whether they will use the proportional price in Exercise 3 or the price developed in Exercise 4.

Bring students together to share and summarize after Exercise 5f.

3 Extend the Activity

Have students use compasses to draw the figure at the right. Then have them find the total area of the shaded region. **Answers depend on the radius. A ≈ 0.54r^2**

4 Assess the Activity

Exercise 5f can be used as an assessment. The report should be assessed on the basis of mathematics used in the report and the completeness of the report.

4-3 Area and Concentric Circles

1 Motivate and Focus

Ask students to describe the effects of dropping an object into a body of calm water. Lead students into a discussion of the definition of concentric circles.

2 Share and Summarize

Bring students together after Exercise 1c to develop a consensus on the formula for finding the length of the rope each day. Summarize again after Exercise 3.

3 Extend the Activity

Have students research the size of an acre. **An acre is 4,840 square yards or 43,560 square feet.** Then ask them to draw a blueprint for the design of the five acres of land described in the activity.

4 Assess the Investigation

Have students complete the research above and determine the possible dimensions the five acres could have. Then have them determine how many days before the goats' grazing areas would overlap if they were tethered to adjacent corners of the property and could not cross the property lines.

Exploring Angles in Circles

● **Mathematical Overview**

The activities in this investigation emphasize the relationships among angles in circles. Three of the four activities involve the use of LOGO. LOGO software uses simple commands to move a drawing cursor called a turtle around the screen. Unless told to do otherwise, the turtle leaves a trail behind it as it is instructed to move.

The most common LOGO software is for Apple computers. However, there are software packages available for Macintosh and IBM compatible machines. The commands given in this investigation are based on Apple Terrapin LOGO and may differ slightly from those in other software programs.

● **Investigation Outcomes**

Students will

- describe the relationship between angles in a circle;
- use computers to investigate circles and stars;
- determine the interior and exterior angles of stars.

● **Materials Needed**

 LOGO software

 straightedge

 compass

 protractor

 local maps

 Blackline Master 5-4

 Transparency Masters 5-1, 5-3

Activity 5-1   The Total Trip Theorem

1 Motivate and Focus

Have each student draw three or four different angles on a piece of paper, using the same corner of the paper as the vertex for all the angles. Ask them how they might find the measures of the angles.

2 Share and Summarize

Have students work in pairs or in appropriately sized groups for the number of computers available. Load the LOGO software. Instruct students to make sure the CAPS LOCK KEY is pressed. Then have students read the first three paragraphs. Use Transparency Master 5-1 to provide a step-by-step introduction to LOGO.

Bring students together after Exercise 1c and 2c to share and summarize.

3 Extend the Activity

Students could be asked to simulate the course they diagrammed in Exercise 1 using LOGO.

4 Assess the Activity

Assessment should be ongoing to judge direction and pace of the instruction and to clarify students' thinking and evaluate their mathematical growth.

Regularly question students to assess their understanding and to help them clarify their thinking. This should be the environment of the class. Students should question each other, and the teacher should question students individually, within small groups, as well as with the entire class. Modification of the direction and pace of the curriculum should follow the assessment.

Students write an explanation about the total trip theorem. The paragraph should be assessed on the basis of the mathematics the student uses in the explanation and the completeness of the explanation.

Activity 5-2 Creating Regular Polygons

1 Motivate and Focus

Allow time for students to experiment with LOGO commands. Then ask students to determine how many turtle steps (units) it takes to get from the bottom edge of the screen to the top edge of the screen. **The screen is approximately 238 units from top to bottom.**

2 Share and Summarize

This activity should move along quickly. Whole class discussion of discovered properties should be conducted after Exercise 5b.

3 Extend the Activity

Students could explore other LOGO commands such as REPEAT, PU, and PD. Then they could be asked to write a procedure using them. Other extension activities are available in the book *101 Ideas for LOGO*, Terrapin Software.

4 Assess the Activity

Exercise 5b can be used as an assessment. The report should be assessed on the mathematics used in the report and the completeness of the report.

Teaching Suggestions and Strategies

Activity 5-3 Inscribed Angles

1 Motivate and Focus

Have students cut a rubber band and tape the ends to cardboard. Have them stretch the rubber band to form an angle, mark the vertex, and sketch the angle formed. Have students measure the angle and one of the legs of the angle. Then have them stretch the rubber band along the leg that was measured until it is twice as long as it was before. Have students measure the newly formed angle and compare it to the original angle. Relate this to inscribed angles in circles.

2 Share and Summarize

Following group discussion of Exercise 1e, bring students together to share their findings. You may want to have them compile their results on Transparency Master 5-3. Bring the class together again after Exercise 2c to discuss the results of their tested conjectures.

3 Extend the Activity

Give students the following problem.

In the circle at the right, inscribed quadrilateral $ABCD$ has the given arc measures. Find the ratio of m$\angle A$ to m$\angle B$.
$$\frac{10x + 14}{9x - 2}$$

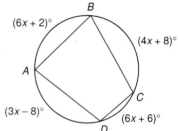

4 Assess the Activity

Give each student a different arc measure.
Have students use a protractor and a compass to draw a circle with three inscribed angles that intercept an arc of the given measure.

Activity 5-4 Inscribed Stars

1 Motivate and Focus

Use Transparency Master 5-1 for those students who may have been absent for previous activities or to review LOGO commands.

2 Share and Summarize

You may want to provide a copy of Blackline Master 5-4 for students to record their work on Exercises 7 a-f. Bring students together after Exercise 7g to discuss the methods they used in completing parts a-f.

3 Extend the Activity

Students may write additional LOGO programs to investigate angular measures and geometric principles.

4 Assess the Investigation

Ask students, "How do you think a LOGO program might be like a game of Treasure Hunt where clues are left at each location?"

Have students write in their journals the completion to the following statement: "The part I liked best about working with the computer is ..."

Ask students, "What mathematics do you now know that you did not know at the beginning of the investigation?"

Teacher's Answer Key

Page 2 1b. Sample answer: The distance from the center to a vertex of each figure should not be more than 4 cm.

Page 3 2c. (The viewing window is [0, 10] by [0, 30] with a scale factor of 1 on both axes.)

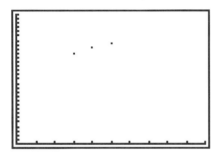

Page 4 4a. (The viewing window is [0, 15] by [0, 30] with a scale factor of 1 on both axes.)

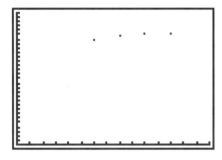

Page 5 5c. (The viewing window is [0, 60] by [0, 40] with a scale factor of 2 on both axes.)

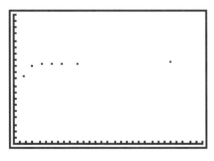

5d. Answers will vary. Sample answer: Perimeter has a limit. The greater the number of sides, the closer it will look to a circle.

Page 6 3e. Sample answer: The ratio between the circumference and diameter is 3.14. The slope of the line is π or approximately 3.14. This shows that no matter how large a circle, the ratio of the distance around the pool to the distance across the pool will be the same.

Page 7 3f. Sample answer: The ratio between the distance across the pool and the cost of building the perimeter of the pool will always be the same. **4.** Answers will vary. Sample answer: Circle—easily adapted to small yards, often used for shallow wading pools, but can be deep enough for diving; Rectangle—best for competitive swimming, available in several construction materials; Teardrop—fits into most gardens; Kidney—curves can be modified to fit the particular site, works with most landscaping; L-shape—fits easily into a corner or around a house projection, diving and swimming areas are defined by the pool shape; Freeform—best for crowded yards or irregular areas.

Page 8 2b. Sample answer: There is about 1.9 inches of space between the two rings. The diameter of the larger ring is always about 3.8 inches longer than the diameter of the smaller ring. **3.** Sample answer: The company can either cut the new rings and recycle the excess or use them and redesign the rest of the materials needed to make the barrels. If the new rings are used, the diameter of the new barrels will all be about 3.8 inches longer. If all other factors remain constant, the most dramatic increase in size will be for the smaller sizes of barrels. For example, if an old barrel had a circumference of 9 inches, the new barrel would be 445% larger. If an old barrel had a circumference of 36 inches, the new barrel would be about 78% larger.

Page 9 4. U.S. ring sizes arbitrarily start with an inner circumference of about 39 mm for a size 1 ring. The circumference increases about 2.5 mm for each whole size increase. A size 8 ring would have a circumference of about 56.6 mm. The measurements for the graph below were rounded to the nearest tenth. (The viewing window is [0, 15] by [10, 30] with a scale factor of 1 on both axes.)

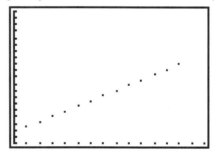

5. Metal is soldered onto the ring or cut out of it to size it. Jewelers have tables to help them determine the length of the metal blank needed for various styles and sizes of rings. The thickness of the metal used affects the actual length needed, since when the blank is formed into a ring shape, the inner surface is compressed and the outer surface is stretched.

Page 10 2b. The diameter is increasing as the number of drops increases, but at a slower rate **2c.** The circumference is also increasing, but at a decreasing rate.

Teacher's Answer Key **T27**

Page 11 4. Sample answer: As the number of drops increases, the diameters also increase, but at a slower rate. At the beginning, the diameter increases a lot with each drop but with a larger number of drops, the diameter does not increase very much with each new drop. **5.** Rainfall is measured using an instrument called a rain gauge which is placed on the ground away from buildings and trees. The rain gauge is cylindrical and contains a smaller cylindrical tube that has a diameter $\frac{1}{10}$ the diameter of the larger cylinder. The smaller tube is attached to a funnel into which rain flows. A ten-inch reading on the smaller tube means that 1 inch of rain has fallen.

Page 13 1a.

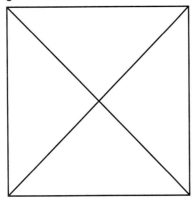

1b. The four triangles are all the same size. The area of one triangle is one-fourth the area of the square. **2.** Sample answer: Yes; Since all regular polygons have a center, they can be separated into triangles using the same procedure.
3a. Sample answer: about 27 cm²; $A = 0.5asn$ where a is the apothem, s is the length of a side of the polygon, and n is the number of small triangles around the center of the polygon.
Page 14 5c. Sample answer: The area would be about 314.2 cm². Since the 100-sided polygon looks like a circle, estimate the circumference or perimeter by finding the product of 3.14 and the diameter and then multiplying that product by the product of 0.5 and the radius, or apothem. **6.** Answers will vary. Sample answer: Yes, the formula for regular polygons would work for circles. The formula $A = \pi r^2$ can be broken up into $A = (r)(\pi r) = (0.5r)(2\pi r) = (0.5r)(C) = 0.5rC$. Since r is the same as apothem and C is the perimeter of the circle, $A = 0.5rC = 0.5ap$. Therefore, the two formulas can be used interchangeably.

Page 15 1b. Sample answer:

length (yd)	12	20	24	30	40
width (yd)	10	6	5	4	3

Page 15 1e. (The viewing window is [0, 40] by [0, 90] with a scale factor of 10 on both axes.)

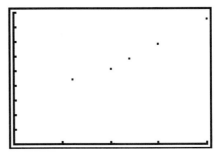

1f. Sample answer: If the dimensions are whole numbers, a corral that is 10 yards by 12 yards uses the least amount of fence—44 yards.

Page 16 3e. Sample answer: A square (or dimensions closest to a square) will get the greatest area for a fixed perimeter.

Page 18 1. Both $3\frac{1}{7}$ and $3\frac{10}{71}$ are correct to 2 decimal places. Estimates may vary.

Page 19 3a.

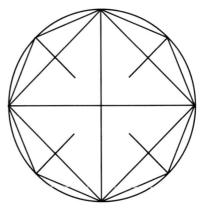

Sample answer: All the sides of the octagon are the same because they were formed by bisecting all the sides of a square. **3b.** Sample answer: $3\frac{3}{64}$ or about 3.05; Multiply the length of one side of the octagon by 8. **4c.** Sample answer: 3.1; about $3\frac{1}{8}$ or 3.125;

Draw some sides, measure several and take an average of these. Then multiply by 32.

Page 20 4f. Answers will vary. Sample estimate 128 sides. Sample answer: Archimedes' circle was much larger so he could be more accurate.

Page 21 3a. $\frac{1}{4}\pi$; Because there is an infinite number of points and one could not throw an infinite number of darts. Every spot would have to be hit once.

Page 22 5b. If the value of d is less than or equal to 1, the dart landed in or on the quarter circle. If the value of d is greater than 1 it is in or on the part of the square that's not covered by the quarter circle.

Page 26   1a. $\frac{355}{113} \approx 3.1415929$,

$\frac{62832}{20000} = \frac{3927}{1250} = 3.1416$,

$\frac{377}{120} = 3.1416667$, $\sqrt{10} \approx 3.1622777$

1b. Answers will vary. Sample answer: Using the graphing calculator is more efficient and the more darts that are thrown the more accurate the approximation.

2b. Sample programs for the infinite series and infinite product approximations for pi:

Prgm1: INFINSER
:Disp "NUMBER OF TERMS"
:Input T
:3 → P
:1 → N
:If T = 1
:Goto 2
:Lbl 1
:1 + N → N
:P + (-1)^N * (1/(2*N+1)) → P
:Disp "π IS APPROX.", P
:If T > N
:Goto 1
:Stop

Prgm2: INFPROD
:Disp "NUMBER OF TERMS"
:Input T
:2 → P
:1 → D
:2 → N
:If T = 1
:Goto 3
:Lbl 1
:P * (N/D) → P
:2 + D → D
:Disp "π IS APPROX.", P
:T − 1 → T
:If T = 1
:Goto 2
:P * (N/D) → P
:2 + N → N
:Disp "π IS APPROX.", P
:T − 1 → T
:If T = 1
:Goto 2
:Goto 1
:Lbl 2
:Stop
:Lbl 3
:Disp "π IS APPROX.", P
:Stop

2c. The infinite series is accurate to one decimal place for 3 terms. The infinite product is accurate to one decimal place for 17 terms. The infinite series doesn't get any more accurate. The infinite product must combine 1439 places to be accurate to three decimal places.

Page 28   4. If you know the degree measure of an angle and you need to find the radian measure, multiply the number of degrees by $\frac{\pi}{180}$. If you know the radian measure and you need to find the degree measure, multiply the number of radians by $\frac{180}{\pi}$.

Page 30   5b. Each slice will be 5° smaller: 8 slices are 45° each and 9 slices are 40° each.

Each slice will have an arc length of $\frac{1}{6}\pi$ in. smaller since the arc length for 8 slices is $\frac{3}{2}\pi$ in. each and for 9 slices it is $\frac{4}{3}\pi$ in. Each slice will have an area 0.5π in² smaller: 8 slices are 4.5π in² each and 9 slices are 4π in² each.

5c. The price of 9 slices of a large pizza sold individually would cost $12.15. This is 27 cents more than the proportional price from Exercise 3.   **5d.** If each pizza is cut into 8 slices, a slice of a large pizza has the same number of degrees but a greater arc length than a regular pizza. A slice of a large pizza is about 4.32 square inches larger than a slice of a regular pizza.
5e. The regular size pizza costs a customer the most per square inch so it will result in the greatest profit.   **6.** If a sector of a circle has an area of A square units, a central angle measuring $N°$, and a radius of r units, then $A = \frac{N}{360}\pi r^2$.   **7.** If the jumbo pizza is cut into 9 pieces the area of the pan pizza should be about 44.7 square inches. This would be a square pan about 6.7 inches on each side or a circular pan with a radius of about 3.8 inches.

However, if the jumbo pizza is cut into 8 pieces, the area of the pan pizza should be about 50.3 in². This would be a square pan about 7.1 in. on each side or a circular pan with a radius of about 4 in.   **8.** Answers will vary. Sample answer: The area of a sector of a circle is a fractional part of the area of the whole circle and is therefore equal to the area of the circle times the fractional part of the circle, or $\frac{N°}{360°} \cdot \pi r^2$.

day	1	2	3	4	5	6	7
rope length (yd)	1.78	2.52	3.09	3.57	3.99	4.37	4.72

Page 31   1c.
Sample answer: Multiply the number of the day by 10, divide by π, and find the square root of the quotient.

Page 32   3b. Sample answer: doubling the area increases the radius by about 41%. Increasing the area by multiples of the original area increases the radius by two-thirds of the previous increase.   **5.** Answers will vary. Sample answer: Area changes by the ratio of the radii of the two circles.

Page 34   2b. The sum is always a multiple of 45°.

Page 35   1c. Sample answer: A figure resembling a circle was drawn. The program drew a figure with sides 1 unit long and with arc measures of 1° for 360°.   **2b.** Sample answer: A figure resembling a semi-circle was drawn. The program drew a figure with sides 1 unit long and with arc measures of 1° for 180°. This figure is half of the previous figure. The last variable made the difference.   **3b.** Sample answer: An octagon was drawn. The program drew a figure with sides 20 units long and with arc measures of 45° for 360°. The measure of each interior angle is 135°.   **3d.** Sample answer: A triangle was drawn. Dividing the last variable by the second variable is the number of sides. The measure of each interior angle is 60°. Subtract the second variable from 180°.   **4b.** Sample answer: A regular hexagon was drawn. Divide the last variable by the second variable to

Teacher's Answer Key   **T29**

find the number of sides. **4d.** Sample answer: A square was drawn twice. The 720° makes the turtle go around twice. The measure of each interior angle is 90°. Reasons will vary.

Page 36 1c. An angle is an inscribed angle if its vertex lies on a circle and its sides contain chords of the circle. The measure of $\angle AXB$ is one-half the measure of $\angle AOB$.

Page 37 1d. $\angle AYB$ is an inscribed angle. $m\angle AYB = \frac{1}{2}m\angle AOB$; $m\angle AYB = m\angle AXB$

1e. Sample answer: The measure of an inscribed angle is one-half the measure of its intercepted arc. The measure of inscribed angles that intercept the same arc are congruent.
2a. Answers will vary. Sample answer. The measure of a central angle is twice the measure of its corresponding inscribed angle. Reasons for conjecture will vary.

Page 38 1b. Sample answer: Object requires 1 more input. There is no condition command to stop Infinite. **2b.** A pentagon was drawn. Sample answer: The program drew a figure with sides 40 units long and with arc measures of 72°. The measure of each interior angle is 108(. Subtract 72° from 180°. **2d.** A 5-point star is drawn. Sample answer: They both have 5 vertices. The arc measure is different. It is a star because 144 does not divide into 360 evenly. The measure of each interior angle is 36°. Subtract 144° from 180°.
3b. Sample answer: The program drew a figure with sides 30 units long and with arc measures of 40°. The figure is a 9-sided regular polygon. The measure of each interior angle is 140°. Subtract 40° from 180°.

Page 39 3d. A 9-point star is drawn. Sample answer: The program draws a figure having sides that are the length of the first variable and having each interior angle measure the difference of 180 and the second variable. **4b.** A septagon is drawn. Sample answer: The program drew a figure with sides 30 units long and with arc measures of 51.43°. The measure of each interior angle is 128.57°. Subtract 51.43° from 180°.
4d. A 7-point star is drawn. Sample answer: They both have 7 vertices. The arc measure is different. The rational number does not have any significant effect. **5a.** Sample answer: A figure with sides 30 units long and with arc measures of 36° will be drawn. **5b.** A decagon is drawn. The measure of each interior angle is 144°. Subtract 36° from 180°. **5c.** Sample answer: A figure with sides 30 units long and with arc measures of 108° will be drawn. Answers will vary. **5d.** A 10-point star is drawn. They both have 10 vertices. The last variable of the second figure drawn is twice as large as the last variable of the first figure drawn. The last variable of Exercise 5c is three times as large as the last variable of Exercise 5a.

Page 40 7a. 75, 144 **7b.** 50, 108 **7c.** 50, 100 **7d.** 50, 170

Page 41 7e. 50, 160 **7f.** 40, 80 **7g.** Sample answer: The value of the side is 43% larger than the length in millimeters. To find the value of the exterior angle measure, find the intercepted arc and divide that by 2 to find the interior angle. The exterior angle, or turning angle, is the supplement of the interior angle. For example, the 36-pointed star has an intercepted arc of 20°. So the interior angle was 10° and the turning angle is 170°.

Teacher Notes

Teacher Notes

Exploring Circles

An Alternative Unit for Analyzing Circles

Author
David Foster

GLENCOE
Mathematics Replacement Units

GLENCOE
McGraw-Hill

New York, New York
Columbus, Ohio
Mission Hills, California
Peoria, Illinois

Copyright ©1996 by Glencoe/McGraw-Hill. All rights reserved.
Except as permitted under the United States Copyrights Act, no part of this publication may be reproduced or distributed in any form or by any means, or stored in a database or retrieval system, without prior written permission of the publisher.

Printed in the United States of America.

Send all inquiries to:
Glencoe/McGraw-Hill
936 Eastwind Drive
Westerville, Ohio 43081

ISBN: 0-02-824218-1(Student Edition)
ISBN: 0-02-824220-3 (Teacher's Annotated Edition)

1 2 3 4 5 6 7 8 9 10 VH/LH-P 04 03 02 01 00 99 98 97 96 95

NEW DIRECTIONS IN THE MATHEMATICS CURRICULUM

Exploring Circles is a replacement unit developed to provide an alternative to the traditional method of presentation of selected topics in Pre-Algebra, Algebra 1, Geometry, and Algebra 2.

The NCTM Board of Directors' Statement on Algebra says,

> "Making algebra count for everyone will take sustained commitment, time, and resources on the part of every school district. As a start, it is recommended that local districts—... 3. experiment with replacement units specifically designed to make algebra accessible to a broader student population."
> (May, 1994 *NCTM News Bulletin*.)

This unit uses analysis of circles as a context to introduce and connect broadly useful ideas in geometry and algebra. It is organized around multi-day lessons called investigations. Each investigation consists of several related activities designed to be completed by students working together in cooperative groups. The focus of the unit is on the development of mathematical thinking and communication. Students should have access to computers with LOGO software and/or calculators with the capability to produce graphs.

About the Author

David Foster received his B.A. in mathematics from San Diego State University and has taken graduate courses in computer science at San Jose State University. He has taught mathematics and computer science for nineteen years at the middle school, high school, and college level. Mr. Foster is a founding member of the California Mathematics Project Advisory Committee and was Co-Director of the Santa Clara Valley Mathematics Project. Mr. Foster is a member of many professional organizations including the National Council of Teachers of Mathematics and regularly conducts in-service workshops for teachers. He is also the author of 13 of Glencoe's *Interactive Mathematics: Activities and Investigations* units.

CONSULTANTS

Each of the Consultants read all five investigations. They gave suggestions for improving the Student Edition and the Teaching Suggestions and Strategies in the Teacher's Annotated Edition.

Linda Herald Bowers
Mathematics Teacher
Alcorn Central High School
Glen, Mississippi

William Collins
Mathematics Teacher
James Lick High School
San Jose, California

Louise Petermann
Mathematics Curriculum Coordinator
Anchorage School District
Anchorage, Alaska

Dianne Pors
Mathematics Curriculum Coordinator
Eastside Union High School
San Jose, California

Javier Solorzano
Mathematics Teacher
South El Monte High School
South El Monte, California

Table of Contents

To The Student   1

Investigation 1   Exploring Circumference

Activity 1–1	Perimeters of Regular Polygons	2
Activity 1–2	The Relationship of Circumference and Diameter	5
Activity 1–3	More on Circumference and Diameter Relationship	7
Activity 1–4	Circles and Functions	9

Investigation 2   Exploring Area

Activity 2–1	Area of Regular Polygons	12
Activity 2–2	Maximizing Area	14
Activity 2–3	Circles and Quadrilaterals	17

Investigation 3   Exploring π

Activity 3–1	Inscribing Regular Polygons	18
Activity 3–2	Using Coordinate Geometry to Calculate π	20
Activity 3–3	More Methods that Generate π	25

Investigation 4   Exploring Arcs and Sectors

Activity 4–1	Arcs and Radians	27
Activity 4–2	The Area of Circles and Sectors	29
Activity 4–3	Area and Concentric Circles	31

Investigation 5   Exploring Angles in Circles

Activity 5–1	The Total Trip Theorem	33
Activity 5–2	Creating Regular Polygons	35
Activity 5–3	Inscribed Agnles	36
Activity 5–4	Inscribed Stars	37

Graphing Calculator Activities

Activity 1	Plotting Points	42
Activity 2	Tables	43
Activity 3	Perimeters of Inscribed Polygons	44
Activity 4	Constructing Lines	45

To the Student

The most often asked question in mathematics classes must be "When am I ever going to use this?" One of the major purposes of *Exploring Circles* is to provide you with a positive answer to this question.

There are several characteristics that this unit has that you may have not experienced before. Some of those characteristics are described below.

Investigations   *Exploring Circles* consists of five investigations. Each investigation has one, two, or three related activities. After a class discussion introduces an investigation or activity, you will probably be asked to work cooperatively with other students in small groups as you gather data, look for patterns, and make conjectures.

Projects   A project is a long-term activity that may involve gathering and analyzing data. You will complete some projects with a group, some with a partner, and some as homework.

Portfolio Assessment   These suggest when to select and store some of your completed work in your portfolio.

Share and Summarize   These headings suggest that your class discuss the results found by different groups. This discussion can lead to a better understanding of key ideas. If your point of view is different, be prepared to defend it.

Exploring Circumference

Circular objects are very common in our world. Circles have unique qualities that make them very useful. Studying shapes such as circles and understanding their attributes is a part of mathematics called *geometry*. Geometry is used by architects to design structures. Using space and materials effectively and efficiently are aspects of building designs that use the principles of geometry. The following activities are examples of how geometry is used to solve problems in the real-world situations.

Activity 1-1 Perimeters of Regular Polygons

Materials

 ruler

 graph paper

 graphing calculator

A new arena is going to be constructed at the local university. A study is being done to find the best performance or playing area design. Since the arena will be used for many different sports, as well as shows and concerts, the designers want a seating arrangement that allows spectators to be as close as possible to the action. They also want to seat as many spectators as possible. The designers decide to investigate several different-shaped seating arrangements. Their goal is to have no front row seat more than 20 meters from the center of the performance or playing area. The arena should also seat as many spectators as possible around the performance area.

● GROUP PROJECT 1

Your task is to investigate several different shapes of figures that have the following specifications.

- The farthest distance from a front row seat to the center of the figure is 20 meters.
- The distance around the figure is as long as possible.

1. Begin by looking at simple closed figures, called **polygons**, that can be made by the front row seats. Look at the figures on Blackline Master 1-1A. Notice that in each polygon all the sides are congruent and all the angles are congruent. When this is true, the polygon is called a **regular polygon**.

   a. The center of a polygon is found at the intersection of the perpendicular bisectors of two adjacent sides of the polygon. Locate the center of each polygon on Blackline Master 1-1A. **See students' work.**

   b. The shapes on Blackline Master 1-1A are **scale drawings** of the possible front row designs of the arena. The scale is determined by the ratio of the length on the drawing to its corresponding length in reality. The designers used a scale of 1 cm = 5 m. How can you be sure the drawings meet the specification that no front row seat is more than 20 meters from the center? **See the Teacher's Answer Key.**

c. The **perimeter** of a geometric figure is the sum of the measures of all its sides. Find the perimeter of each figure on Blackline Master 1-1A.

d. The table below shows the relationship of the number of sides of each shape to its perimeter in centimeters. Copy and complete the table.

Number of sides	3	4	5
Perimeter (cm)	**20.8**	**22.6**	**23.5**

Share & Summarize

e. **Journal Entry** What relationship do you notice in the table? Be prepared to share your findings with the class. **Sample answer: As the number of sides increases, the perimeter increases.**

● HOMEWORK PROJECT 1

Graphing Calculator Activity

You can learn how to use a graphing calculator to plot points in Activity 1 on page 42.

2. a. On a piece of graph paper, draw a large L to represent the horizontal and vertical axes. Write *Number of Sides* along the horizontal axis and *Perimeter (cm)* along the vertical axis. **See students' work.**

b. Number the tick marks on each axis to make a scale appropriate for the data in Exercise 1d. **See students' work.**

c. Plot each point (*number of sides, perimeter*) using the pairs of values from the table in Exercise 1d. **See the Teacher's Answer Key.**

d. Do the points appear to have a pattern? If so, describe it. **Yes, they appear to lie on a line.**

e. What might the graph look like with several more points plotted? **Answers will vary. Sample answer: a parabola.**

f. What conjectures can you make about the number of sides in a figure and its perimeter? **Answers will vary.**

g. What would you recommend to the designers of the arena at this time? **Answers will vary. Sample answer: Use the pentagon design.**

● GROUP PROJECT 2

3. a. Repeat the procedure you used in Exercises 1a and 1c for the regular hexagon (6 sides), regular octagon (8 sides), regular decagon (10 sides), and regular dodecagon (12 sides) on Blackline Master 1-1B. Remember to make sure that the center of each figure is 20 meters from the farthest point on the perimeter of the figure. **See students' work.**

b. The table below shows the relationship of the number of sides of each shape to its corresponding perimeter in centimeters. Copy and complete the table.

Number of sides	6	8	10	12
Perimeter (cm)	**24**	**24.8**	**25**	**25.2**

Investigation 1 **Exploring Circumference**

Share & Summarize

c. Discuss the data in your table with the class. What relationship do you notice in the table? Did the pattern you found in the first table continue in the second table? **Answers will vary. The general pattern of perimeter increasing as the number of sides increases continues.**

HOMEWORK PROJECT 2

4. a. Plot each point (*number of sides, perimeter*) on a coordinate plane using the values from the table in Exercise 3b. **See the Teacher's Answer Key.**

   b. Do the points appear to have a pattern? If so, describe it. **Sample answer: The last three points appear to be on a line.**

   c. Is this graph consistent with the conjectures you made in Exercises 2d and 2f? Explain how you have either supported or revised your conjecture. **See students' work.**

   d. Generalize a relationship between the number of sides of regular figures with congruent distances from the center to each vertex point and their perimeters. **Sample answer: As the number of sides increases so does the perimeter.**

   e. What would you recommend to the designers of the arena at this time? **Sample answer: Use the dodecagon design.**

5. The graphing calculator program below uses trigonometry to determine the length of a side of any regular polygon with a radius (distance from the center of the figure to each vertex point) of 4 centimeters and the polygon's perimeter. Although you may not understand the mathematics of the program at this time, the program itself is useful in investigating the perimeters of regular polygons.

The program is written for use on a TI-82 graphing calculator. If you have a different type of programmable calculator, consult your User's Guide to adapt the program for use on your calculator.

```
PROGRAM: SIDELENG
:Degree
:Disp "NUMBER OF SIDES"
:Input S
:(180/S) → A
:(8*sin A) → L
:S*L → P
:Disp "LENGTH"
:Disp L
:Disp "PERIMETER"
:Disp P
:Stop
```

   a. To find the perimeter of any regular polygon, first access the program, and press ENTER . Then enter the number of sides of your polygon, and press ENTER . The screen will display the length of one side and the perimeter of the polygon.

4   Exploring Circles

b. Using the program, test your generalization on polygons having various numbers of sides. Keep track of the number of sides and the perimeter. Be sure to try polygons with a large number of sides.

c. Arrange the data you collected in a table and display the data on a coordinate plane either by drawing a graph or using a graphing calculator. **See the Teacher's Answer Key.**

d. Will the perimeters of the polygons keep getting larger? Is there a limit to the number of sides a polygon can have? Explain why or why not. **See the Teacher's Answer Key.**

e. What would you recommend to the designers of the arena at this time? Explain your conclusion in detail. Be prepared to explain your reasoning to the class. **See students' work.**

● HOMEWORK PROJECT 3

6. Extension The performance area of an arena may come in many different shapes. Select three or four shapes and explain the advantages and disadvantages of differently-shaped performance areas in an arena. **See students' work.**

Activity 1-2 The Relationship of Circumference and Diameter

You are a manager for an above ground swimming pool company. Your company makes circular pools that have a plastic lining. The sides of the pool are supported by a wooden structure, and the pool includes a short deck and fence around the outside of the pool. The wooden structure, deck, and fence are more costly than the plastic liner. Some customers are interested in pools in which they can swim long laps across the pool. Of course, these same customers are interested in the lowest price possible. Your manager thinks that some size pools are better than others for these customers. In other words, he believes certain circles have more distance across the circle in relation to the distance around the circle. He has asked you to investigate this matter and report back to him on which pool has the best ratio of distance around the pool to distance across the pool.

Investigation 1 **Exploring Circumference**

● GROUP PROJECT

Materials

 cylindrical objects

 tape measure

Share & Summarize

1. The **circumference** of a circle is the distance around the circle. The distance across the circle through its center is called its **diameter**.

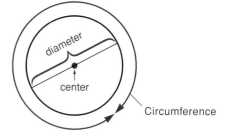

   a. Locate at least six cylindrical objects of various sizes around your classroom. Make a table and record the measurement of the diameter and the corresponding circumference of each object. **See students' work.**

   b. What relationship do you notice in the table? **Sample answer: The circumference is about 3 times the diameter.**

2. a. Plot each point (*diameter, circumference*) on a coordinate plane using the pairs of values from the table in Exercise 1a. **See students' work.**

   b. Do the points appear to have a pattern? If so, describe it. **Sample answer: They appear to be in a line.**

   c. What conjectures can you make about the measurement of the diameter and the corresponding circumference of each circular object? **Sample answer: The circumference is about 3 times the diameter.**

   d. What would you recommend to your manager about his theory? What size pool gives you the longest laps for the lowest price? Be prepared to share your reasoning with the class. **Answers will vary. Sample answer: His theory does not hold true**

3. a. Draw a line between two different points that you plotted in Exercise 2. How steep is the line? **Sample answer: It is very steep.**

   b. In mathematics, the steepness of a line is called its **slope**. One way to find the slope of a line is to examine the vertical and horizontal change between two points on the line. The slope is the ratio of the vertical change to the horizontal change. To determine the slope of the line in Exercise 3a, locate two points on the line. Write their coordinates. Subtract the first coordinates from each other to find the change in x. Subtract the second coordinates from each other to find the change in y. Divide the change in y by the change in x. What did you calculate for the slope? **Answers will vary but should be about 3.**

   c. Compare your answers to other students' answers in your class.

   d. Add another column to the table you made in Exercise 1a. Divide the circumference of each circle by its diameter. Write the answer in the new column. **See students' work.**

   e. What relationship do you notice about the values in the new column? How do these relate to the slope of the line? What does this information tell you about your investigation? **See the Teacher's Answer Key.**

6 Exploring Circles

Share & Summarize

f. What would you recommend to your manager about his theory? What size of the pools gives you the longest laps for the lowest price? Write an explanation to your manager with your findings. Use mathematics to defend your explanation. Be prepared to share your reasoning with the class. **See the Teacher's Answer Key.**

Journal

g. Journal Entry Explain what you know about the relationship of the diameter of the circle to the circumference. **Sample answer: The circumference is about 3 times the diameter.**

HOMEWORK PROJECT

4. Extension Swimming pools come in many different shapes. List some of the different shapes that are often used by pool makers. Give a possible reason as to why pools are designed in the shapes they are and list the advantages and disadvantages of each differently-shaped pool. **See the Teacher's Answer Key.**

Activity 1-3 More on Circumference and Diameter Relationship

You are a quality controller for a barrel manufacturing company. Your company makes barrels of all different sizes. Some barrels are used by farmers for storing crops. Some of your smaller barrels are cut in half and used as buckets. You make very large storage vats for vineyards. You even make very small barrels as earrings that you sell in the gift shop next to your factory. Without exception you make every barrel the same way. You use wood supported with metal rings that go around the barrels.

Your ring-making machine is operated by a computer. Last week the computer malfunctioned and made every ring, from the smallest for the earrings to the largest for the wine vats, longer than normal by exactly one foot. In other words, every ring your factory has turned out recently has a circumference that is twelve inches longer than the correct size.

FYI The barrel for most liquids holds about $31\frac{1}{2}$ gallons, or 119 liters.

Materials

 cylindrical objects

 tape measure

 string

 scissors

GROUP PROJECT

1. Your supervisor has asked you to investigate this matter. What should the company do with all those rings? It would be very costly to dispose of them. Your supervisor has asked you to tell her what size barrels need to be made to use those rings. Are the barrels usable? How would the size of the new barrels compare to the size of the original barrels?

Investigation 1 **Exploring Circumference**   **7**

 **Journal**

a. **Journal Entry** Before you begin to investigate this problem, write down your initial thoughts. How will the new rings compare to the old rings? How much bigger will the new rings look in comparison to the barrels that they were supposed to fit? How will the new rings compare to the small barrels, like the earrings? How will the new rings compare to the large barrels, like the wine vats? **See students' work.**

b. Select four or five cylindrical objects ranging in size from very small (like a barrel the size of an earring) to very large (like the top of a wine vat) to some in between. For a very large circle, use string and chalk to draw a circle on a large area such as a parking lot.

c. Use some string and measure the circumference of one of the cylindrical objects. Cut the string and then cut another piece that is twelve inches longer. Tape the two ends of the shorter piece of string together to form a ring. Then tape the two ends of the longer piece of string to form a second ring. **See students' work.**

d. Compare the two rings. How much bigger is the second ring than the first? Place the larger ring around the smaller ring so that a smaller circle is centered inside a larger circle. These are called *concentric circles*. How much space is there between the two circles? How much longer is the diameter of the larger ring than that of the smaller ring? **12 in.; about 1.9 in.; about 3.8 in.**

2. a. Repeat the process in Exercise 1 for the other three or four cylindrical objects you selected. Be sure your objects differ in size. **See students' work.**

   b. For each object, answer the following questions.

   - How do the circumferences of the smaller and larger rings compare? **The larger are always 12 inches longer.**
   - How much space is there between the two rings? **See the Teacher's Answer Key.**
   - How much longer is the diameter of the larger ring than that of the smaller ring? **See the Teacher's Answer Key.**

 **Share & Summarize**

3. Write a summary report that explains what you have discovered. Be sure that your summary answers the following questions. Be prepared to share your reasoning with the class. **See the Teacher's Answer Key.**

   - What will you recommend to your supervisor in regards to the use of the new rings?
   - If the company uses the new rings, how will it affect the size of new barrels? Will the new barrels be usable?
   - What is the relationship between the size of the new barrels and the old barrels?

4. **Extension** Rings that you wear on your fingers come in different sizes. Someone might wear a size 8 ring. What does that measurement mean? How is it related to the diameter or circumference of the ring itself? Research ring sizes. Go to the library or talk with a jeweler. Make a graph showing the relationship of ring sizes to the diameter and/or the circumference of a ring. **See the Teacher's Answer Key.**

5. **Extension** If a ring that you wear on your finger doesn't fit, you can have the ring sized. Talk with a jeweler and research the process that is performed in sizing a ring. How does the jeweler know how much to add or subtract from the circumference of the ring to make it the right size **See the Teacher's Answer Key.**

Activity 1-4 Circles and Functions

You work for the municipal water district in the water conservation department. You have decided to investigate the effect of dripping water faucets. You want to find out the amount of water that is lost by a dripping faucet. You have decided to set up an experiment to see how much water is lost in relationship to the number of drips from a faucet in a certain time period. With this information, you can estimate the amount of lost water from hundreds of faucets and use the information to mount a water conservation campaign.

Materials

 eyedropper

 paper towels

 ruler

 colored pencils

 graph paper

GROUP PROJECT

1. **a.** Lay down a paper towel. Mark a small point at the center of the paper towel. Fill the eyedropper with water. Hold the eyedropper about two inches above the point marked on the paper towel. Squeeze out one drop of water from the eyedropper. Measure the diameter of the circular water spot on the paper. **See students' work.**

    b. Find the circumference of the water spot in part a. The circumference of a circle is equal to its diameter times π, which is about 3.14. **See students' work.**

Investigation 1 **Exploring Circumference**   9

c. Record the diameter and circumference of the water spot in the first row of a table like the one shown at the right.
See students' work.

Number of Drops	Diameter	Circumference
1		
2		
3		
4		
5		

d. After you have recorded and calculated the diameter and circumference of one drop, repeat the process by squeezing a second water drop from the eyedropper onto the center point of the paper towel. Measure the diameter of the spot now to see how much it expanded. Calculate the new circumference and complete the second row of the chart.
See students' work.

e. Before you repeat the process, look at your table. Write down what you expect if you continue to squeeze drops on the mark on the paper towel. Estimate how large the water spot will be in terms of the length of the diameter if you were to drip 100 drops onto the paper towel. Estimate how many drops it might take for part of the water spot to reach a side of the paper towel. **See students' work.**

Share & Summarize

f. Continue to squeeze drops of water from the eyedropper on the center point, measuring the diameter each time and calculating the circumference. Record all your data in the table. Do this for at least twelve drops. **See students' work.**

g. Examine the data in your table. Compare the data with your estimates in part e. Be prepared to share your results with the class.
See students' work.

2. a. Using the table you made in Exercise 1, make a graph using two different sets of points on the same axes. Let the horizontal axis of your graph represent the number of drops and the vertical axis represent the number of centimeters. Using a colored pencil, plot each point (*number of drops, diameter*). Then, using a different colored pencil, plot (*number of drops, circumference*). **See students' work.**

b. What is happening to the diameter of the spot as the number of drops increases? Does the diameter increase at the same rate after each drop? Estimate what the diameter will be after 20 drops. **See the Teacher's Answer Key.**

c. What is happening to the circumference as the number of drops increases? Estimate what the circumference will be after 20 drops.
See the Teacher's Answer Key.

d. What is the relationship between the set of diameter points and the set of corresponding circumference points? **Sample answer: The circumference points are about 3 times higher than the corresponding diameter points.**

10   Exploring Circles

 e. Check your estimate for the diameter and circumference of the spot after 20 drops by starting with a new towel, especially if the paper has started to dry. Squeeze 20 drops onto a paper towel at the center point. Measure the diameter and calculate the circumference.
See students' work.

 f. How do the actual measures compare with your estimate? Are you able to predict, with any kind of accuracy, the length of the diameter and circumference given a certain number of drops? Generalize a formula or process for finding the diameter. **Answers will vary.**

Share & Summarize

3. How does this experiment help you with the original problem? What have you discovered about dripping water? What techniques, predictions, or estimates can you use to make sense of the original problem? Write a summary of your conclusions. Be prepared to share your reasoning with the class. **See students' work.**

Journal

4. **Journal Entry** The relationship between the number of drops and the size of the circle is a function. Explain what you know about the function.
See the Teacher's Answer Key.

 HOMEWORK PROJECT

Portfolio Assessment

A portfolio is repesentative samples of your work, collected over a period of time. Begin your portfolio by selecting an item that shows something new you learned in this Investigation.

5. **Extension** Rainfall is recorded in inches. A weather person might report that a certain region has recorded 15 inches of rain so far this year. How is that calculated? Research the method used to find the rainfall for a region. What does rainfall inches mean in terms of diameter, circumference, and volume? **See the Teacher's Answer Key.**

Investigation 1 **Exploring Circumference**   **11**

Exploring Area

Finding the area of a shape has many real-world applications. The relationship between the area and other measures of a geometric figure can provide you with a better mathematical understanding of and insight into the figure. In this investigation, you will see connections between the area of circles and relevant applications. You will also develop generalizations and formulas about the area of geometric figures. This is an important skill in mathematics.

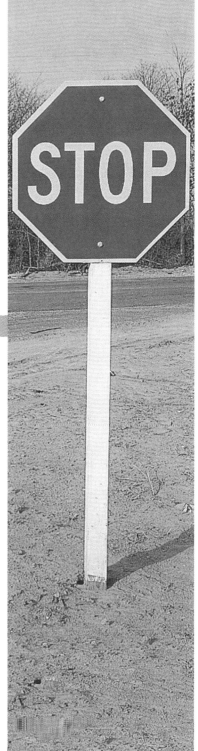

Activity 2-1   Area of Regular Polygons

Materials

centimeter grid paper

The Department of Transportation has decided to use differently-shaped symbols to communicate information to motorists. They have decided to use regular polygons because they are recognizable and pleasing to view. They already use octagons for stop signs, equilateral triangles for yield signs, and squares for "do not enter" signs. They have decided to use pentagons for "one way" signs, hexagons for "wrong way" signs, and decagons for general information signs.

● **PARTNER PROJECT**

1. Your supervisor at the Department of Transportation has given you the task of determining the amount of material and paint needed to make these signs. The supervisor asked you to investigate the possibility of a single method for finding the area of a regular polygon so you wouldn't have to use a separate formula for each figure.

12   Exploring Circles

a. You remembered that separating a shape into smaller simpler shapes is a good problem-solving strategy to use when finding its area. You also remembered that all regular polygons have a center point that is the same distance from all of the vertices. Since a triangle is the simplest polygon in terms of the number of sides, you separate a regular polygon into triangles. Each small triangle uses the center point of the polygon as one vertex and the other two vertices are vertices of one side of the polygon. On centimeter grid paper, draw a square with sides 5 centimeters long, and separate it into four small triangles.
See the Teacher's Answer Key.

b. How do the four triangles compare? What is the relationship between the area of one triangle and the area of the entire square?
See the Teacher's Answer Key.

c. The formula for the area (A) of a triangle is $A = \frac{1}{2}bh$ where b is the base and h is the height. Find the area of one of the small triangles you formed in part a. **6.25 cm²**

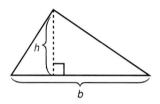

d. Using your answer to part b, find the area of the square. Explain how you calculated your answer. **25 cm²; 4 times 6.25 cm²**

Share & Summarize

e. Find the area of the square using a different method. How does the area compare to your answer in part d? **25 cm²; They are the same.**

2. Do you think the technique you used in Exercise 1 to find the area of a square will work for all polygons? Explain your reasoning. **See the Teacher's Answer Key.**

3. a. A segment that is drawn from the center of a regular polygon perpendicular to a side of the polygon is called an **apothem**. On Blackline Master 2-1, find the center of the pentagon using the method from Activity 1-1. Then draw small triangles so that the center is one vertex and the other two vertices are vertices of one side of the pentagon. Measure a side of the pentagon and the height of a small triangle (apothem) and then calculate the area of the pentagon. Explain in writing how you calculated your answer. Write a formula that represents how you found the area. **See the Teacher's Answer Key.**

b. Use the formula you developed in part a to find the area of the other regular polygons on Blackline Master 2-1. **about 29 cm²; about 33.6 cm²**

Investigation 2 **Exploring Area of Circles** 13

4. **a.** What is the relationship between the number of small congruent triangles around the center of a regular polygon and the number of sides of the polygon?   **The number of triangles is equal to the number of sides.**

   b. What do you get when you multiply the number of sides of a regular polygon times the length of a side?   **perimeter of the polygon**

   c. How can the perimeter of a regular polygon be used in your formula for finding the area of the polygon?   **It can replace s (length of a side) times n (number of small triangles).**

   d. Rewrite your formula to involve the perimeter. Explain why the formula works.   **A = 0.5ap where p represents the perimeter; The formula works because s times n equals the perimeter.**

Share & Summarize

   e. Refer to the original problem where your supervisor wanted you to determine the amount of material and paint needed to make a pentagonal, a hexagonal, and a decagonal sign. Your supervisor also wants to use one method for finding the area of these signs. How does what you discovered about the area of a polygon help you with the problem? Write an explanation to your supervisor explaining your findings. Be prepared to share your reasoning with the class. **See students' work.**

HOMEWORK PROJECT

5. Since every regular polygon has a center, a circle can be drawn so that the vertices of the polygon are on the circle. Thus, the distance from each vertex to the center is the radius of the circle.

   a. Suppose you had to calculate the area of a regular polygon with 100 sides. If the radius of the polygon is 10 centimeters, what would that polygon look like?   **Sample answer: It would look almost like a circle with a diameter of 20 centimeters.**

   b. Why could you use the formula for the circumference of a circle to estimate the perimeter of the 100-sided polygon?   **Sample answer: Because a 100-sided polygon is almost a circle.**

   c. Use the formula for the circumference of a circle to estimate the area of the 100-sided regular polygon. Explain how you arrived at your estimate.   **See the Teacher's Answer Key.**

6. **Extension** Your supervisor advised you that the Department of Transportation was instituting the symbol of a circle to distinguish important landmarks. She asked if your formula for regular polygons would work for circles. She wanted to know if you could explain the relationship of the formula of the area of a circle and the formula for the area of a regular polygon. Prepare a report for your supervisor.   **See the Teacher's Answer Key.**

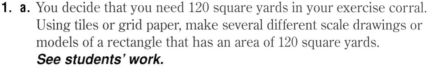

Activity 2-2 Maximizing Area

You are a rancher and you raise quarter horses. You need to construct a new exercise corral on your ranch. You have a limited budget for fencing and would like to construct an exercise corral with the largest area using the least amount of fence. A neighbor rancher suggested building a rectangular exercise corral. You decide to investigate what dimensions would be best for the exercise corral.

FYI
Quarter horses acquired their name from the fact that they were originally used to race on a straight quarter-mile track.

GROUP PROJECT

Materials

 graph paper

 tiles

 calculator

1. **a.** You decide that you need 120 square yards in your exercise corral. Using tiles or grid paper, make several different scale drawings or models of a rectangle that has an area of 120 square yards. **See students' work.**
   b. Make a table showing the length and the width of each rectangle you drew or modeled in part a. **See the Teacher's Answer Key.**
   c. Find the perimeter of each rectangle in part a. Make a list of the perimeters from least to greatest. **See students' work.**
   d. Share your list with the rest of the class. Are there dimensions of rectangles not on the list whose area is 120 square yards? Were numbers other than whole numbers used to find dimensions? **Answers will vary.**
   e. Using a graphing calculator or graph paper, construct a graph of the data you created. Let the horizontal axis represent the length of one side of a rectangle, and let the vertical axis represent the length of the corresponding perimeter. **See the Teacher's Answer Key.**

Graphing Calculator Activity
You can learn how to use a graphing calculator to make a table and sort lists in Activity 2 on page 43.

Share & Summarize

   f. Write an explanation of what size exercise corral you would build that has an area of 120 square yards and uses the least amount of fence possible. Be prepared to share your reasoning with the class. **See the Teacher's Answer Key.**

2. Another neighbor rancher gave you a great idea for your exercise corral. She told you to use one side of your barn as one of the sides of the corral. Your barn is 15 yards wide.

   a. Using this side of your barn as one of the sides of the corral, how much fence would you need to build an exercise corral that has an area of 120 square yards and uses the least amount of fence? What would be the dimensions of the corral? **31 yards; The dimensions are 8 yd by 15 yd.**

Share & Summarize

   b. How many different dimensions for the corral did you consider? Why did you select the corral you did? Be prepared to share your work with the class. **See students' work.**

Investigation 2 **Exploring Area of Circles**   **15**

PARTNER PROJECT

3. a. At the hardware store, you discover that fencing costs $12 for one yard. What would be the cost of the corral you designed without using a side of the barn? What is the cost for fence if you use a side of the barn? How much did you save? **about $526; about $372; about $154**

    b. Suppose you were already prepared to spend the money on the original design. How big of a corral, in terms of area, could you make if you bought the same amount of fence and used a side of the barn? What would be the dimensions of this corral? **about 216 yd^2; 15 yd by 14.4 yd**

    c. You decide to save some of the money you were going to spend in building an exercise corral. You are going to use the side of the barn, but you're not sure how much fence to use. What dimensions should you use to get the most area for the least cost of the fence? How much fence would you need to buy? What is the cost-to-area ratio? **See students' work.**

Share & Summarize

    d. Write a report summarizing your findings. Explain how you arrived at your conclusions and what process you used. Give specific examples to support your findings. Illustrate your findings with graphs, diagrams, or other pictorial representations. Be prepared to discuss your report with the class. **See students' work.**

Journal

    e. **Journal Entry** Describe what you know about finding the greatest rectangular area when the perimeter stays the same. **Answers will vary.**

GROUP PROJECT

4. While you were at the hardware store researching fence prices, you explained your idea to the sales clerk and she told you that using the barn wasn't such a good idea. First, the location of the barn is too close to your house. The horseflies and noise might be disturbing. She also said that the horses will kick in the side of the barn and it could be more expensive in the long run. She said you should reconsider your plan for the corral. She suggested a design different from a rectangle.

    a. Investigate other geometric shapes for a corral and determine which shape would give you the most area for the least price. Remember you don't want to use a side of the barn. Make a list of possible shapes along with the areas and length of fence surrounding those areas. **See students' work.**

Share & Summarize

    b. You decide you can afford to spend $265 on fencing. What shape will enclose the largest area possible? What are the dimensions of the shape? Draw a blueprint of the shape, label its dimensions, and write an explanation of your solution. Be prepared to share your solution with the class. **Sample answer: A circular region with a diameter of 7 yards.**

16 Exploring Circles

Activity 2-3 Circles and Quadrilaterals

Materials

 graph paper

 pennies

AA Aluminum is a company that manufactures aluminum cans. They are able to make the sides of the cans with little or no waste. However, there is a lot of waste in cutting out the tops from sheets of aluminum. The manufacturing manager has asked you to investigate the problem. Your task is to find a way to arrange the patterns of circular tops on the sheet of aluminum in order to cut the most tops out of the aluminum sheet. The diameter of the top of a can is 6 centimeters. The sheet of aluminum is a square 60 centimeters on each side.

GROUP PROJECT

1. **a.** Make a scale drawing of the top of a can by tracing a penny. Measure the diameter of the tracing. Now make a scale drawing of the square sheet of aluminum. How big should you make the square so that it is the same scale? **See students' work.; 20 cm on each side**

    b. Make scale drawings of three or four patterns. **See students' work.**

    c. Determine the number of tops generated from each pattern in part b. Find the amount of waste and compare it to the amount of usable aluminum. **See students' work.**

Share & Summarize

   d. Select the pattern with the least amount of waste and justify the merits of the pattern. Determine the number of tops produced from a sheet and how much waste was left over. What percentage of the material was usable? Explain the geometry of your pattern and why it uses space efficiently. Explain why you believe it is the best pattern. Be prepared to share your findings with the class. **See students' work.**

HOMEWORK PROJECT

Portfolio Assessment

Select one of the assignments from this Investigation that you found especially challenging and place it in your portfolio.

2. **Extension** After you made your report to AA Aluminum, the manufacturing manager asked if there was a better-shaped figure to use other than the square sheet of aluminum. The manager also wanted to know what the best size or dimensions of the sheet of aluminum would be. Investigate shapes other than squares and find out if another shape would allow less waste. What are the dimensions of the shape that will produce the most tops with the least amount of waste? **See students' work.**

Investigation 2 **Exploring Area of Circles**   **17**

Exploring π

Archimedes is considered one of the greatest mathematicians of all time. He lived in the Greek city of Syracuse, Sicily, between 287 B.C. and 212 B.C. and is known for his work in many areas of mathematics and science.

Activity 3-1   Inscribing Regular Polygons

Materials

 poster paper

 string

 meter stick

 ribbon

Archimedes' mathematical work includes inventing methods of calculating the area and volume of geometric objects, such as circles, cylinders, cones, spheres, and ellipses. He also spent time investigating the decimal approximation of π, the ratio of the circumference of a circle to its diameter. The method he used to approximate π will be similar to the method you will use in this activity. Archimedes, as the story is told, drew a large circle on the floor of his home. He measured the diameter of the circle and called that distance one unit. Then he inscribed different regular polygons inside the circle and measured each perimeter. As the number of sides of the polygons increased, the shapes of the polygons looked more like circles, and the perimeters approached the value of π. Using this method, he was able to calculate an approximation of π correct to a value between $3\frac{1}{7}$ and $3\frac{10}{71}$.

● GROUP PROJECT

1. How precise was Archimedes' approximation of π? Give your answer in terms of the number of correct decimal places. Estimate how many sides his final polygon needed to be to get a number accurate to that number of decimal places.   **See the Teacher's Answer Key.**

2. Following Archimedes' method, start this activity with a large piece of butcher or poster paper. The paper should be slightly longer and wider than a meter.

   a. Mark a point in the center of the poster paper. Using the point as the center, construct the largest circle possible. You may need to use a pencil attached to a string. Next draw a diameter of the circle. If the length of the diameter is one unit, what is the measure of the circumference?   **See students' work; π.**

   b. Cut a piece of ribbon that is longer than the diameter of the circle in part a. Using the diameter of the circle to represent one unit, make a homemade tape measure out of the ribbon.   **See students' work.**

18   Exploring Circles

c. Draw a second diameter perpendicular to the first diameter. Explain the method you used and how you are sure your two diameters are perpendicular to each other. **Sample answer: The measures of the central angles are each 90°.**

d. Inscribe a square in the circle by connecting the points where the diameters intersect the circle. Measure the perimeter of the square using your ribbon tape measure. Make a table and record the number of sides of the inscribed polygon and the corresponding perimeter. **See students' work.**

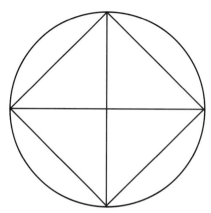

3. a. Construct the perpendicular bisector of each side of the square. The perpendicular bisectors will intersect the circle at four points. Connect each new point to the vertices of the square that are on either side to form an octagon. Explain how you know all the sides of the octagon are the same length. **See the Teacher's Answer Key.**

  b. Using your "tape measure," determine the perimeter of the octagon and record the new data in the table you made in Exercise 2d. Explain how you calculated the perimeter. **See the Teacher's Answer Key.**

4. a. Inscribe a 16-sided regular polygon in the circle by repeating the process you used in Exercise 3a. Construct the perpendicular bisector of each side of the octagon, and connect the points where they intersect the circle to the vertices of the octagon. Describe the relationship between the 16-sided polygon and the circle. **Sample answer: The sides of the polygon are closer to the circle.**

  b. Determine the perimeter of the 16-sided polygon and record that information in your table. Use the information in the table to construct a graph. Plot each point (*number of sides, perimeter*) using the pairs of values from the table. Describe the graph once it has been drawn. **See students' work. Answers may vary.**

Share & Summarize

  c. Estimate the perimeter of a 32-sided regular polygon if it was inscribed in the circle. Find the perimeter without drawing all the sides of the polygon. Instead find the perimeter by drawing only one side. Explain how you were able to find the perimeter. **See the Teacher's Answer Key.**

  d. Complete your table with the results of your calculations from the 32-sided polygon inscribed in the circle. How accurate is your approximation of π at this point? How close is your number to the actual value of π? **Answers will vary.**

Investigation 3 **Exploring** π

Graphing Calculator Activity

You can learn how to use a graphing calculator to find the perimeter of an inscribed polygon in Activity 3 on page 44.

Share & Summarize

e. Continue this process of finding the perimeter of regular polygons inscribed in a circle as a method for approximating π. What is the greatest number of sides that you can accurately draw and measure on your paper? What is the perimeter of that polygon? Record the results in your table and graph all your data. Describe your graph. **Answers will vary.**

f. Recall that Archimedes used his method to calculate the approximation of π correct to a value between $3\frac{1}{7}$ and $3\frac{10}{71}$. How precise was his approximation of π in relationship to your own approximation? Give your answer in terms of the number of correct decimal places. Estimate how many sides his final polygon needed to be to get a number accurate to that number of decimal places. How did your effort compare with Archimedes? **See the Teacher's Answer Key.**

g. Write a summary report about finding the value of π using this historical method. Be prepared to share the figure that you have constructed and the table of data that you have recorded with the class. **See students' work.**

Activity 3-2   Using Coordinate Geometry to Calculate π

Materials

centimeter grid paper

compass

blank transparency

bobby pins

In the 17th century, French mathematician René Descartes developed a concept that would forever link algebra and geometry. He developed **coordinate geometry**, which uses ordered pairs of numbers to locate points on a plane. The plane is defined by two perpendicular number lines. This is sometimes referred to as the coordinate plane or Cartesian plane, named after the mathematician. Using this system, mathematicians could explore geometry using algebraic notation. In this activity, you will use this system to explore circles and another method for approximating the value of π.

The center of the coordinate plane, the origin (O), is at 0 on each of the perpendicular number lines. The **ordered pair** for the origin is (0,0). Every point on the plane can be located using an ordered pair.

In the figure at the right, point A is named by the ordered pair (1, 2). The first number, 1, called the *x*-**coordinate**, indicates the number of units to move left or right from the origin. The second number, 2, called the *y*-**coordinate**, indicates the number of units to move up or down from the origin.

20   Exploring Circles

PARTNER PROJECT 1

1. Draw perpendicular number lines on graph paper to form a coordinate plane. Then graph each of the following points on a coordinate plane.

   $A(1, 1)$ $B(-2, 1)$ $C(-2, -2)$ $D(1, -2)$ $E\left(\frac{1}{2}, \frac{1}{3}\right)$
   See students' work.

2. You can use a coordinate system to set up a simulation to approximate the value of π.

   a. Use a compass to draw one quarter of a circle with a radius of 1 unit on Blackline Master 3-2A. **See students' work.**

   b. Enclose the quarter circle by drawing a square with sides 1 unit long. The area of the square is 1 square unit. The area of the quarter circle is found using the formula for the area of a circle ($A = \pi r^2$) and then dividing by four. What is the area of the quarter circle?

   $\frac{1}{4}\pi$ **or** $\frac{\pi}{4}$ **or about 0.785 square units**

   Suppose the square with the quarter circle was a dartboard and you threw one thousand darts at that board. Of course, you would never miss the dartboard completely. So all the points on the dartboard, even points on the square, would have the same chance of being hit by a dart. Every time a dart lands inside or on the quarter circle you would make a tally. You would also count the number of darts you threw. What could be done with the information you would obtain from this simulation? How might it help approximate the value of π?

   You could use the information to write the ratio of the number of darts that landed inside or on the quarter circle to the total number of darts thrown. How might that ratio help us calculate the value of π?

Share & Summarize

3. **a.** What is the ratio of the area of the quarter circle to the area of the square? The ratio of the number of darts that land inside or on the quarter circle to the total number of darts thrown should approach the ratio of the areas. Why do we say *approach* rather than *equal*? What would have to happen for it to be *equal* instead of *approach*? Could that ever happen? Be prepared to share your reasoning with the class.
   See the Teacher's Answer Key.

   b. How can you find an approximation for π from the ratio of the number of darts that land inside or on the quarter circle to the total number of darts thrown? **multiply by 4**

Investigation 3 **Exploring** π

HOMEWORK PROJECT

4. Suppose (x, y) names a point where a dart lands. If you form a right triangle by drawing a vertical line through (x, y), you can use the Pythagorean theorem to find the distance between (x, y) and the origin. The distance, d, is given by

   $$d = \sqrt{x^2 + y^2}.$$

   How far from the origin is the point $(0.9, 0.8)$? **about 1.2 units**

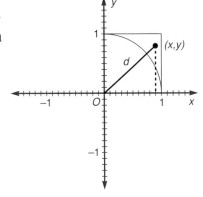

5. **a.** What do you know about a dart if it lands less than or equal to 1 unit from the origin? Explain. **It is on or inside the quarter circle because the radius of the circle is 1.**

   b. How can you use the distance formula in Exercise 4 to tell if a dart lands in or on the quarter circle, or in or on the part of the square that's not covered by the quarter circle? **See the Teacher's Answer Key.**

   c. Write an ordered pair for a point that would be inside the quarter circle. **Sample answer: (0.5, 0.5)**

   d. Write an ordered pair for a point that would be in the square, but not the quarter circle **Sample answer: (0.8, 0.7)**

GROUP PROJECT

You can simulate the throwing of darts by using spinners to randomly generate the location of each dart thrown on our board.

6. **a.** Cut out the sectored circles on Blackline Master 3-2B. Assign the values 0.1, 0.2, 0.3, ..., 0.9 to each region. You can make each sectored circle into a spinner. Place the closed end of a bobby pin over the center of the sectored circle. Then place a pencil or the metal end of a compass in the center and spin the bobby pin. Let one spinner represent the *x*-coordinate, and let the other spinner represent the *y*-coordinate. You may wish to label them. **See students' work.**

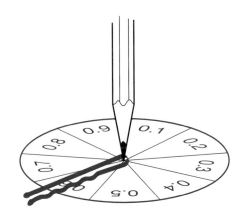

b. On Blackline Master 3-2A, label the tick marks like the illustration shown at the right. This will be your "dartboard."
See students' work.

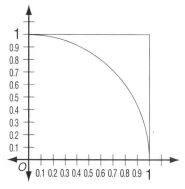

c. Tape a blank transparency over the dartboard. Use a transparency marker to locate each "throw." Have one group member spin the x-coordinate spinner, another group member spin the y-coordinate spinner, a third group member plot points on the dartboard, and a fourth group member keep a running tally of the number of darts that lie on or inside the quarter circle and the number of darts that lie outside the circle. As a group, simulate 50 throws. **See students' work.**

d. After you have completed your simulation, use your data to approximate the value of π. Discuss, with your group, factors that may affect the accuracy of your value of π.
See students' work.

e. Combine the totals of the entire class. Approximate the value of π from the data of the whole class. **See students' work.**

f. Write a summary of your findings. Describe the accuracy of this method of approximating π. Be prepared to share your summary with the class. **See students' work.**

Share & Summarize

PARTNER PROJECT 2

7. The graphing calculator program below also simulates throwing a dart at a square and quarter circle. The program uses a random number generator to insure that each dart has an equal chance of hitting any point in the square.

The program is written for use on a TI-82 graphing calculator. If you have a different type of programmable calculator, consult your User's Guide to adapt the program for use on your calculator.

Note: This program will take 3-4 minutes to run.

```
PROGRAM:DARTS
:ClrHome
:ClrDraw
:0 → C
:0 → N
:Disp "1 VALUE"
:Disp "2 GRAPH"
:Input M
:If M = 1
:Goto 1
:Line(0,1,1,1)
:Line(1,0,1,1)
:DrawF √(1 − X²)
:Pause
:DispGraph
:Lbl 1
:rand→X
:rand→Y
:If M = 2
:PT − On(X,Y)
:√(X²+Y²) →D
:If D ≤ 1
:C + 1 →C
:N + 1 →N
:4*C/N →P
:If M = 1
:Disp P
:If N < 1000
:Goto 1
:If M = 2
:Pause
:Disp "PI IS"
:Disp "APPROXIMATELY"
:Disp P
:Stop
```

a. Before running the program, set the range. The viewing window should be [0, 1.8] by [0, 1.2] with a scale factor of 0.1 on both axes.

b. The program will calculate the distance each dart is from the origin and approximate π using the ratio $\dfrac{\text{number of darts on or inside the quarter circle}}{\text{total number of darts thrown}}$.

To run the program, first access it and press ENTER.

Press 1 ENTER to only perform the calculations with points, and not graph them.

Press 2 ENTER if you want to see the points graphed as they are generated. If you choose the second option, the program first draws the dartboard. Press ENTER to begin throwing the darts. When 1,000 darts have been thrown, the line in the upper right corner will stop flashing. Press ENTER again for the approximation for π using the ratio. **See students' work.**

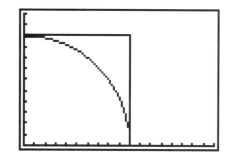

c. How could you modify the program to simulate throwing 2,000 darts?
change If N<1000 to If N<2000

d. Run the program several times and for at least six different total numbers of darts. Record the number of darts thrown and the corresponding approximations for π. **See students' work.**

Share & Summarize

e. Write a report summarizing your findings in this activity. Be sure that your summary answers the following questions.

- Is using technology more valid than doing the simulation by hand?
- Describe the accuracy of this method for approximating π.

Give specific examples to support your findings. Be prepared to discuss your report with the class. **See students' work.**

Activity 3-3 More Methods that Generate π

Materials
 graphing calculator

Throughout history, different approximations for π have been calculated and used. In ancient Asia, the value of π was 3, and in ancient Egypt, the value was calculated to be $\left(\dfrac{4}{3}\right)^4$ or 3.16049. Archimedes' method was the first scientific attempt to calculate π. He approximated it to be between 3.140845 and 3.142857.

Investigation 3 **Exploring** π **25**

GROUP PROJECT

1. A chronological listing of π is shown on the timeline below.

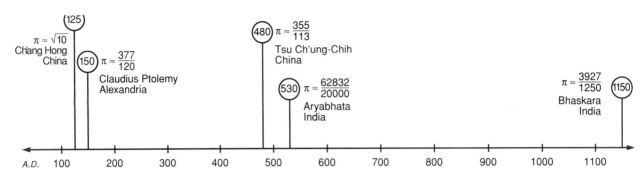

   a. Find the decimal approximation of π for each rational expression on the timeline. Then order the values from most precise to least precise. **See the Teacher's Answer Key.**
   b. How do these approximations compare to the approximations you found using the graphing calculator program in Activity 3-2? Explain which method is most efficient and which is most accurate. **See the Teacher's Answer Key.**

PARTNER PROJECT

Portfolio Assessment

Select an item from this investigation that you feel shows your best work and place it in your portfolio. Explain why you selected it.

2. In the 16th century, the first of several infinite series was found to approximate π. An **infinite series** is the indicated sum of an infinite sequence of terms. The more terms you combine, the more accurate the approximation. An **infinite product** is similar to an infinite series. Instead of adding terms, the terms of an infinite product are multiplied together. Two examples that approximate π are given below.

$$\pi = \frac{2 \cdot 2 \cdot 2 \cdot 4 \cdot 4 \cdot 6 \cdot 6 \cdot 8 \ldots}{1 \cdot 1 \cdot 3 \cdot 3 \cdot 5 \cdot 5 \cdot 7 \cdot 7 \ldots} \qquad \pi = 3 + \frac{1}{3} - \frac{1}{5} + \frac{1}{7} - \frac{1}{9} + \frac{1}{11} - \ldots$$

   a. Which of the two equations gives a more accurate approximation of π? **the infinite series**
   b. **Extension** Write a graphing calculator program to approximate π using one of the two infinite equations above. Write out the listing of your program and explain how it works. **See the Teacher's Answer Key.**
   c. How accurate is your value of π using the graphing calculator program? How many terms must be combined to be accurate to three decimal places? **See the Teacher's Answer Key.**

Share & Summarize

   d. Compare your program to two other programs written by your classmates. Determine how the programs compare in terms of efficiency, accuracy, and speed. How do these programs compare to finding π using the graphing calculator program in Activity 3-2? Explain which method is most efficient and which is most accurate. **See students' work.**

Exploring Arcs and Sectors

Geometric figures are often used to enhance the beauty of architectural design. In the photograph at the right, the stained glass dome is composed of several circles that have the same center. When circles lie in the same plane, have radii of different lengths, and have the same center, they are called **concentric circles**. In this investigation, you will analyze a situation involving concentric circles after you explore other connections between parts of circles.

Activity 4-1   Arcs and Radians

Materials

 pipe cleaners

 scissors

 centimeter ruler

Do the spokes of a wheel have a relationship to the distance around the wheel? Does the relationship change as the size of the wheel changes? You are going to investigate these questions in this activity.

 GROUP PROJECT

1. **a.** Select at least 12 pipe cleaners. Make sure that they are all the same length. Construct a wheel with five or more spokes. What is the ratio of the length of a spoke to the distance around the wheel?   ***about 0.16***

   b. Cut another set of pipe cleaners that are a different length from part a, but each of them of the same new length. Construct a second wheel with five or more spokes. What is the ratio of the length of a spoke to the distance around the new wheel? How does the ratio of the first wheel compare to the ratio of the second wheel?   ***about 0.16; They are the same.***

   c. Cut a third set of pipe cleaners a different length from the first two sets. Make sure the pipe cleaners are all the same length. Construct a third wheel with five or more spokes. What is the ratio of the length of a spoke to the distance around the new wheel? How does the ratio of the third wheel compare to the ratio of the first two wheels?   ***about 0.16; They are all the same.***

Investigation 4 **Exploring Arcs and Sectors**   27

Share & Summarize

d. Does the relationship change as the length of the spoke changes? Generalize the relationship between the length of a spoke of a wheel and the circumference of the wheel. **No; Sample answer: The ratio of the length of a spoke of a wheel to the circumference is about 0.16.**

2. The spokes of the wheels you constructed in Exercise 1 separated the outside of the wheels into **arcs**. An arc is a part of the circle. If the spokes are 1 unit long and the arc between two adjacent spokes is 1 unit long, the measure of the angle formed by the spokes is defined to be 1 **radian**.

   a. What is the circumference of a circle whose radius is 1 unit long? **2π units**

   b. What is the radian measure of one complete revolution of a circle whose radius is 1 unit long? **2π radians**

   c. How many radians make up the length around half of the circle in part b? **π radians**

3. You know that one complete revolution of a circle is 360°. Write a proportion that illustrates the relationship between radians and degrees.

   $$\frac{\text{degree measure}}{360} = \frac{\text{radian measure}}{2\pi}$$

4. Explain how to change the measure of an angle from degrees to radians and from radians to degrees. **See the Teacher's Answer Key.**

5. Angles expressed in radians are often written in terms of π. The term *radians* is also usually omitted when writing angle measures.

   a. Change 90° to radian measure in terms of π. $\frac{\pi}{2}$

   b. Change $\frac{2\pi}{3}$ radians to degree measure. **120°**

   c. How many degrees equal one radian? **1 radian = $\frac{180}{\pi}$ degrees or about 57.3°**

HOMEWORK PROJECT

Share & Summarize

6. **Journal Entry** In your own words, explain the relationship between radians and degrees. Be prepared to share your explanation with the class. **See students' work.**

Activity 4-2 The Area of Circles and Sectors

Materials

calculator

FYI
The first pizza made with tomatoes and cheese was created to match the colors of the Italian flag: red (tomatoes), white (mozzarella cheese), and green (basil).

The pizzeria where you work currently has three sizes of pizzas—regular, large, and jumbo. Their diameters are shown in the illustration below. The store owner has asked you to assist in determining the price for each size of pizza.

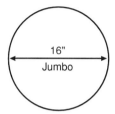

● PARTNER PROJECT

1. The formula for the area of a circle is $A = \pi r^2$.

   a. How much bigger is the jumbo pizza than the large pizza in terms of area?   **about 88 in²**

   b. How much bigger is the large pizza than the regular pizza in terms of area?   **about 34.5 in²**

2. The cheese used to top the pizzas comes in 2-inch square slices. How many more slices of cheese are needed for the jumbo pizza than for the regular pizza?   **Answers will vary. Sample answer: about 31 slices**

3. The price of a regular cheese pizza is $8.25. The owner wants your opinion on what to charge for the large and jumbo size pizzas. He wants to cover the cost of the ingredients and make a proportional amount of profit. What should he charge for a large and jumbo cheese pizza?   **Sample answer: $11.88 for the large and $21.12 for the jumbo**

● HOMEWORK PROJECT

4. The ingredients for the regular cheese pizza cost $2.65. You believe that customers should get a discount on the larger size pizzas. Determine prices for the large and jumbo pizzas that are less than the proportional prices you found in Exercise 3. Explain to the owner how much profit is made by selling the pizzas at your price. Justify your proposal using mathematics.   **See students' work.**

● GROUP PROJECT

Share & Summarize

5. **a.** The owner sells pizza by the slice at lunch. He has been selling a slice of a large pizza for $1.35. He cuts each pizza into eight slices and wants to know whether he will make more money selling by the slice than by selling a whole pizza. Explain how his prices for the slices and whole pizza compare.   **A large pizza sold by the slice costs $10.80. See the Teaching Suggestions.**

Investigation 4 **Exploring Arcs and Sectors**   **29**

Graphing Calculator Activity

You can learn how to use a graphing calculator to find the area of a sector in Activity 4 on page 45.

Share & Summarize

b. The owner has asked you to determine the effect of cutting a large pizza into nine slices. How much smaller will each slice be? Use the terms *degrees*, *arc length*, and *area* in your answer. **See the Teacher's Answer Key.**

c. The owner wants to sell each of the nine slices for the same price as when the pizza was cut into eight slices. Compare the prices of selling the nine slices individually with the price of selling an entire large pizza. Explain how the prices compare. **See the Teacher's Answer Key.**

d. The owner also thinks he can save more money if he sells slices from regular size pizzas instead of from large size pizzas. How much larger is a single slice of a large pizza than that of a slice from a regular pizza? Use the terms *degrees*, *arc length,* and *area* in your answer. **See the Teacher's Answer Key.**

e. The owner has decided to cut his pizzas into nine slices. He is going to sell a regular slice for $1.10, a large slice for $1.35 and a jumbo slice for $1.60. Taking into account that the cost of the ingredients are proportional to the area of the slice, which size will result in the greatest profit? **See the Teacher's Answer Key.**

f. Prepare a report summarizing your group's findings. Illustrate your report with graphs, charts, or other pictorial representations. Be prepared to present your report to the class. **See students' work.**

6. A **sector** of a circle is a region of the circle bounded by a central angle and the intercepted arc. For example, figure *QRTS* is a sector of circle *Q*. You can think of a slice of a round pizza as a model for a sector of a circle. Using what you learned about the relationship between radians and degrees in Activity 4-1 and your findings in Exercise 5, develop a formula for the area of a sector given the measure of the central angle and the radius of the circle. **See the Teacher's Answer Key.**

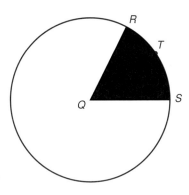

● HOMEWORK PROJECT

7. **Extension** A competitor is selling individual pan pizzas for the same price as two of your jumbo slices. What size would this individual pizza need to be to have the same area as two jumbo slices? **See the Teacher's Answer Key.**

Journal

8. **Journal Entry** Describe the relationship between the area of a sector of a circle and the area of the circle. **See the Teacher's Answer Key.**

Activity 4-3 Area and Concentric Circles

Materials

 calculators

You own five acres of land and have a few animals. You have chickens, rabbits, and a couple of goats. You also have a vegetable garden and some fruit trees. The chickens and rabbits are no problem. The rabbits are fenced off from the garden and never get out. The chickens are well fed and don't bother the garden. The goats on the other hand are a problem. They keep breaking out of their pen, destroying the garden and ruining the fruit trees. You decided that they broke out for the *last* time last week!

GROUP PROJECT

1. **a.** You decide to place a post in the area where the goats graze and tether a goat to it. You know that a goat needs to graze on about ten square yards each day. How long does the rope need to be the first day for a goat to have enough area to graze? **about 1.78 yards**

   b. After the first day, the goat has consumed the grass around the post. You know you must lengthen the rope so that the goat can graze on another 10 square yards. How much longer does the rope need to be the second day? **about 0.74 yards longer**

 **Share & Summarize**

   c. You decide to analyze the data if the rope is lengthened for 7 days to see if this is a good method for containing the goats. Determine the length the rope needs to be each day to insure that the goat has 10 square yards of new grazing area each day. Determine a general formula or process for finding the length of the rope each day. Be prepared to share your findings with the class. **See the Teacher's Answer Key.**

   d. If you put two posts 80 feet apart and tethered a goat to each post, when would the grazing areas overlap each other? **The areas would overlap on the 56th day.**

2. **a.** You have a small rectangular barn. One day you decide to tether a goat to one side of your barn for grazing. The barn is 30 feet long on one side, and you tether the goat to the middle of that side of the barn. How long do you need to make the rope to insure the goat has 10 square yards to graze? **about 2.52 yards**

   b. Your barn is 10 feet wide. If you tethered a goat to the corner of the barn, how long do you need to make the rope to insure it has 10 square yards to graze? **about 2.06 yards**

Investigation 4 **Exploring Arcs and Sectors**

Share & Summarize

3. **a.** Draw conclusions based on your group's findings. Verify your conclusions by using two differently-sized daily grazing areas. ***See students' work.***

   b. After verifying your conclusions, write a summary report that explains what your group discovered. Be sure that your summary answers the following questions.

   - Given a circle, how does doubling the area affect the radius?
   - Given a circle, how does increasing the area by multiples of the original area affect the radius?

   Be prepared to share your report with the class. ***See the Teacher's Answer Key.***

Portfolio Assessment

Select some of your work from this investigation that shows how you used a calculator or computer. Place it in your portfolio.

HOMEWORK PROJECT

4. You decide to tether both goats to the same long side of the barn. How far apart from each other do you need to tether the goats to insure an adequate grazing area for each animal for one day? ***The goats should be about 5.04 yards apart and tethered with a rope 2.52 yards long to assure no overlapping.***

5. **Extension** Describe the relationship between the areas of sectors with equal degree measures in concentric circles. ***See the Teacher's Answer Key.***

Exploring Angles in Circles

Circles and angles are linked together throughout the study of geometry. We measure angles and arcs using degrees or radians. There are many important mathematical relationships between circles and angles that are inscribed in circles, tangents to circles, or intersected by a circle.

In this investigation, the geometric properties of circles and angles will be explored using LOGO. LOGO is a very powerful computer programming language. You will be using its turtle graphics feature. This feature allows you to create a drawing with simple commands.

Once you have LOGO loaded into your computer, you type DRAW to clear the screen and position the turtle, a small triangle used to draw, in the center of the screen. The basic LOGO commands are as follows:

FD	FORWARD	LT	LEFT	CS	CLEARSCREEN
BK	BACK	RT	RIGHT	HT	HIDETURTLE
PU	PENUP	PD	PENDOWN	ST	SHOWTURTLE

Use numbers following direction commands (FD or BK) to indicate the distance you want the turtle to travel. Numbers following turn commands tell the turtle the number of degrees to turn. For example, FD 10 means go forward 10 units, and LT 60 means turn left 60 degrees.

If you use negative numbers, the turtle does the opposite command. For example, RT -30 means turn left 30 degrees, and FD -50 means go backward 50 units.

Clearscreen (CS) clears the graphics screen. Pen-Up (PU) allows you to move the turtle without drawing while Pen-Down (PD) resumes drawing when the turtle moves. Hideturtle (HT) makes the turtle disappear and Showturtle (ST) makes the turtle appear.

Activity 5-1   The Total Trip Theorem

● **PARTNER PROJECT 1**

Materials

 LOGO software

 local map

1. On a map of your local community draw a diagram of a trip from your home to a grocery store, a bank, the library, the post office, and then back to your home.

   a. Measure the angle of each turn on your trip. Assign right turns a positive angle measure and left turns a negative angle measure. Find the sum of all the angle measures in your trip. **See students' work.**

   b. Share the trip you diagrammed with five other pairs of classmates. Record the measures and number of turns each pair took. Compare the sum of the angle measures of each pair of classmates' turns. Describe your findings. **See students' work.**

Share & Summarize

c. What conjectures can you make about the sum of the angle measures in the trips made around your neighborhood? Explain any generalizations that can be found. Be prepared to share your findings with the class. *See students' work.*

● PARTNER PROJECT 2

2. a. Using only the commands FD and RT, write LOGO commands to start at the center, show the turtle and make a path that wanders around the screen and stops at the center. Use various lengths and angles to move around the screen. *Sample answer: DRAW, FD 40, RT 90, FD 80, RT − 90, FD 40, RT − 90, FD 140, RT − 90, FD 80, RT − 90, FD 60*

b. Find the sum of the angle measures that you used in the wandering path in part a. What conjectures can you make about the sum? *See the Teacher's Answer Key.*

Share & Summarize

c. **Journal Entry** What conclusions can you make about a trip that starts and ends at the same location? Write a paragraph summarizing your conclusions. Give specific examples to support your findings. Be prepared to share your summary with the class. *Sample answer: The sum is always between −450° and −270° and between 270° and 450°.*

Activity 5-2 Creating Regular Polygons

In this activity, you will use LOGO to investigate regular polygons and their characteristics or attributes.

● PARTNER PROJECT

Materials

LOGO software

1. a. Enter the procedure below using the editor in LOGO. See the LOGO appendix in the User's Guide for the steps on how to enter and exit the LOGO editor for your computer.

```
TO   OBJECT  :SIDE  :ANGLE  :TOTAL
IF  :TOTAL  <  0  THEN  STOP
FD :SIDE  RT  :ANGLE
OBJECT  :SIDE  :ANGLE  :TOTAL  −  :ANGLE
END
```

b. Once you have entered the procedure above, exit the editor, type the following lines, and press the RETURN key.

```
DRAW
OBJECT  1  1  360
```

c. What happened? How do you think the 1 1 360 relates to the figure that was drawn? Explain how the program works. **See the Teacher's Answer Key.**

2. a. Type the following two lines and press the RETURN key.

   CS
   OBJECT 1 1 180

   b. What happened? How do you think the 1 1 180 relates to the figure that was drawn? How does this figure compare to the figure that was drawn in Exercise 1b? What made the difference in the two figures? **See the Teacher's Answer Key.**

3. a. Type in the following two lines and press the RETURN key.

   DRAW
   OBJECT 20 45 360

   b. What happened? How do you think the 20 45 360 relates to the figure that was drawn? What is the measure of each interior angle of the figure? Explain how you found this measure. **See the Teacher's Answer Key.**

   c. Type in the next two lines and press the RETURN key.

   CS
   OBJECT 30 120 360

   d. What was drawn? How might you know what would be drawn? What is the measure of each interior angle of the figure? Explain how you found this measure. **See the Teacher's Answer Key.**

4. a. Sketch the figure you think will be drawn when you enter the two lines below. Check your prediction by typing the lines and pressing the RETURN key. The turtle will draw a regular hexagon.

   CS
   OBJECT 20 60 360

   b. What was drawn? What information did you use to make your prediction? **See the Teacher's Answer Key.**

   c. Describe what you think will be drawn when you enter the following two lines. Explain your reasoning. Check your prediction by typing the lines and pressing the RETURN key.

   CS
   OBJECT 40 90 720 **The turtle will draw a square. Answers will vary.**

   d. What was drawn? How does the value 720 affect the outcome of the drawing? What is the measure of each interior angle of the figure? Explain your reasoning. **See the Teacher's Answer Key.**

Investigation 5 **Exploring Angles in Circles**

5. **a.** Describe what you think will be drawn when you enter the following two lines. Explain your reasoning. Check your prediction by typing the lines and pressing the RETURN key.

> CS
> OBJECT 1 1 90   ***The turtle will draw a quarter circle.***
> ***Answers will vary.***

Share & Summarize

b. You are asked to write a software manual to explain what you just investigated. Write a report explaining how the procedure OBJECT works. Give examples of different figures that can be created using the procedure. Explain how to draw circles, polygons, and arcs. Be prepared to share your report with the class.   ***See students' work.***

Activity 5-3   Inscribed Angles

In this activity, you will discover the relationships among central angles, inscribed angles, and arcs.

Materials

 compass

 protractor

 straightedge

Radii is the plural of radius.

● **GROUP PROJECT**

1. **a.** Use a compass to draw a circle that has a radius at least 3 inches long. Mark the center of the circle, and label it point O. Draw two radii from the center of the circle to form an angle less than 90 degrees. Label the two points where the radii intersect the circle points A and B. Since the vertex of ∠AOB is at the center of the circle, ∠AOB is called a **central angle**.   ***See students' work.***

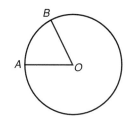

b. We say that ∠AOB *intercepts* arc AB. Arc AB is written $\widehat{AB}$. $\widehat{AB}$ is called the **intercepted arc** of ∠AOB. The measure of $\widehat{AB}$ is the same as the measure of ∠AOB. Use a protractor to find the measure of ∠AOB. Record the measure of ∠AOB.   ***See students' work.***

c. Select a point on circle O that is not on $\widehat{AB}$. Label the point X. Draw $\overline{AX}$ and $\overline{BX}$. Since these segments have their endpoints on the circle, they are called **chords** of the circle. ∠AXB is an **inscribed angle**. Write a definition of an inscribed angle. Then compare the measure of ∠AXB to the measure of ∠AOB.   ***See the Teacher's Answer Key.***

36   Exploring Circles

d. Select another point on circle O that is not on $\overset{\frown}{AB}$. Label the point Y. Draw $\overline{AY}$ and $\overline{BY}$. How would you describe $\angle AYB$? Find the measure of $\angle AYB$. How does the measure of $\angle AYB$ compare to the measure of $\angle AXB$? **See the Teacher's Answer Key.**

Share & Summarize

e. As a class, compile each group's results in a chart. Describe any patterns you notice in the completed chart. **See the Teacher's Answer Key.**

HOMEWORK PROJECT

2. a. Make a conjecture about the relationship between the measures of central angles and the measures of the corresponding inscribed angles. Explain why you believe your conjecture is true. **See the Teacher's Answer Key.**

b. Test your conjecture several times. Keep a record of your tests. **See the Teacher's Answer Key.**

Share & Summarize

c. How do the results of your tests compare with your conjecture? If your tests seem to help validate your conjecture, explain why. If your tests seem to discredit your conjecture, then explain how you might revise your conjecture to conform with the results of the additional tests. Be prepared to share your findings with the class. **See students' work.**

Activity 5-4   Inscribed Stars

Materials

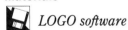
LOGO software

The LOGO procedure OBJECT that you used in Activity 5-2 is known in computer programming as a finite looping procedure. It produced regular polygons when you entered different lengths and angle measures. You will now explore a procedure designed to run indefinitely. The procedure must be stopped by the user after a figure is drawn. To stop the procedure, press the appropriate keys as designated by your computer or LOGO program.

Apple II Terrapin LOGO:	press CTRL-G
IBM or Compatible LCSI LOGO:	press CTRL-BREAK
Macintosh LOGO:	press Command –

Investigation 5 **Exploring Angles in Circles**   **37**

PARTNER PROJECT

1. **a.** Enter the following procedure using the LOGO editor:

   ```
   TO   OBJECT. INFINITE  :SIDE  :ANGLE
   FD   :SIDE RT   :ANGLE
   OBJECT  INFINITE  :SIDE  :ANGLE
   END
   ```

   b. Notice the similarities and differences between the procedure OBJECT and OBJECT.INFINITE. Describe the differences and predict how the execution of the procedures might be different. **See the Teacher's Answer Key.**

2. **a.** Once you have entered the procedure in Exercise 1, exit the editor, type the following lines, and press the RETURN key.

   ```
   DRAW
   OBJECT.INFINITE   40   72
   ```

   Remember, you must manually stop the execution of the procedure.

   b. What happened? How do you think the 40 72 relates to the figure that was drawn? What is the measure of each interior angle of the figure? Explain how you found this measure. **See the Teacher's Answer Key.**

   c. Type the following lines and press the RETURN key.

   ```
   CS
   OBJECT.INFINITE   40   144
   ```

   d. What happened? What is similar about this figure and the figure in part 2a? What made the difference in the two figures? What is the measure of each interior angle of this figure? Explain how you found this measure. **See the Teacher's Answer Key.**

3. **a.** Type the next two lines and press the RETURN key.

   ```
   CS
   OBJECT.INFINITE   30   40
   ```

   b. How do you think the 30 40 relates to the figure that was drawn? How does the figure compare to the figures drawn in Exercises 2a and 2c? What is the measure of each interior angle of this figure? Explain how you found this measure. **See the Teacher's Answer Key.**

   c. Predict what you think will happen when you enter the following two lines and press RETURN. Explain your reasoning. Check your prediction by typing the lines and pressing the RETURN key.

   ```
   CS
   OBJECT.INFINITE   30   80
   ```
   Sample answer: A figure with interior angles that each measure 100° will be drawn.

d. What happened? Describe what you know about the OBJECT.INFINITE procedure. **See the Teacher's Answer Key.**

4. a. Type the following two lines and press the RETURN key.

```
CS
OBJECT.INFINITE  30  51.43
```

b. What happened? How do you think the 30 51.43 relates to the figure that was drawn? What is the measure of each interior angle of the figure? Explain how you found this measure. **See the Teacher's Answer Key.**

c. Type the next two lines and press the RETURN key.

```
CS
OBJECT.INFINITE  30  102.86
```

d. What happened? What is similar about this figure and the figure in part 4a? What made the difference in the two figures? How does a rational number affect the procedure? **See the Teacher's Answer Key.**

5. a. Describe what you think will be drawn when you enter the following two lines. Explain your reasoning. Check your prediction by typing the lines and pressing the RETURN key. **See the Teacher's Answer Key.**

```
CS
OBJECT.INFINITE  30  36
```

b. What happened? What is the measure of each interior angle of the figure? Explain how you found this measure. **See the Teacher's Answer Key.**

c. Describe what you think will be drawn when you enter the following two lines. How will the figure compare to the figure drawn in part 5a? Check your prediction by typing the lines and pressing the RETURN key. **See the Teacher's Answer Key.**

```
CS
OBJECT.INFINITE  30  108
```

d. What happened? What is similar about this figure and the figure in part 5a? What is the relationship between the last variable in each pair of figures in Exercises 2-4? Why is this relationship different in figures drawn in Exercises 5a and 5c? **See the Teacher's Answer Key.**

Journal

6. Journal Entry Explain why the measure of the interior angles helps you predict what the figure will look like. **See students' work.**

GROUP PROJECT

7. The following procedure generates the figures shown below. Your teacher may give you a copy of Blackline Master 5-4, which contains the same figures.

   ```
   TO    WHAT   :SIDE   :ANGLE
   FD    :SIDE
   RT    :ANGLE
   WHAT  :SIDE  :ANGLE
   END
   ```

   Determine the appropriate values for :SIDE and :ANGLES that will create each of these figures. Make predictions and then use LOGO to verify your predictions. Determine a general method for finding the measure of an interior angle of each figure. ***See the Teacher's Answer Key.***

   a.
   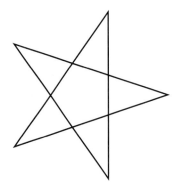

   b.

   c.

   d.

e.

f.

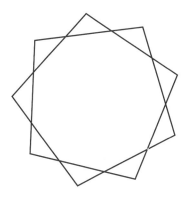

Share & Summarize

g. Discuss your findings with the class. Be prepared to share the method you used for finding the values of the side and angle for each star.
See the Teacher's Answer Key.

HOMEWORK PROJECT

Journal

h. **Journal Entry** Given a star of any number of points, explain how to determine the measure of the exterior and interior angles without using a protractor or other measuring device.

Portfolio Assessment

Select some of your work from this investigation that shows how you used a calculator or computer. Place it in your portfolio.

Investigation 5 **Exploring Angles in Circles**

Graphing Calculator Activities

Graphing Calculator Activity 1: Plotting Points

The graphics screen of a graphing calculator can represent a coordinate plane. The x- and y-axes are shown, and each point on the screen is named by an ordered pair. You can plot points on a graphing calculator just as you do on a coordinate grid.

The program below will plot points on the graphics screen. In order to use the program, you must first enter the program into the calculator's memory. To access the program memory, use the following keystrokes.

Enter: [PRGM] [▶] [▶] [ENTER]

Example Plot the points $(-7, 4)$, $(-3, -1)$, $(5, 4)$, $(1, -6)$, and $(-8, -2)$ on a graphing calculator.

First set the range. The notation $[-12, 12]$ by $[-9, 9]$ means a viewing window in which the values along the x-axis go from -12 to 12 and the values along the y-axis go from -9 to 9.

```
Prgm1: PLOTPTS
:FnOff
:PlotsOff
:ClrDraw
:Lbl 1
:Disp "X="
:Input X
:Disp "Y="
:Input Y
:Pt-On(X, Y)
:Pause
:Disp "PRESS Q TO QUIT,"
:Disp "1 TO PLOT MORE"
:Input A
:If A = 1
:Goto 1
```

The program is written for use on a TI-82 graphing calculator. If you have a different type of programmable calculator, consult your User's Guide to adapt the program for use on your calculator.

Enter: [WINDOW] [ENTER]
[(-)] 12 [ENTER] 12 [ENTER] 1 [ENTER]
[(-)] 9 [ENTER] 9 [ENTER] 1 [ENTER]

Now run the program.

Enter: [PRGM] 1 [ENTER]

Enter the coordinates of each point. They will be graphed as you go. Press [ENTER] after each point is displayed to continue in the program.

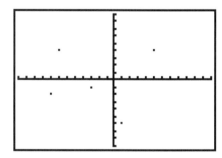

● **Try This**

Use the program to graph each set of points on a graphing calculator. Then sketch the graph on a sheet of graph paper.   *See students' work.*

1. $(-7, 1)$, $(3, -6)$, $(1, -2)$, $(-6, -8)$
2. $(1.7, -2.2)$, $(-0.8, -1.9)$, $(1.2, -0.1)$, $(2.1, 3.7)$, $(-1.6, -3.2)$
3. $(32, -4)$, $(-25, 15)$, $(13, 18)$, $(5, 11)$
4. $(-92, -40)$, $(67, -21)$, $(-51, 37)$, $(-24, 16)$, $(-32, 57)$, $(-89, -21)$

Graphing Calculator Activity 2: Tables

A graphing calculator is a powerful tool for studying functions. One way you can examine a function is to create a table of values. A TI-82 graphing calculator will allow you to create a large table of values quickly.

Example The height in feet, y, of a model rocket after it is launched is represented by the equation $y = 45x - 5x^2$, where x is time elapsed in seconds. Use a graphing calculator to create a table of values for $x = \{0, 0.5, 1, 1.5, 2, 2.5, 3, 3.5, ...\}$. After 4 seconds, how high is the rocket?

First, enter the function $y = 45x - 5x^2$ into the Y= list. Press [Y=] to access the list. Then enter the equation in as function Y_1. Use the [CLEAR] key to remove any equations that are already in the list.

Now press [2nd] [TblSet] to display the table setup menu. The table is to start at 0, so enter 0 as the TblMin value and press [ENTER]. ΔTbl is the change between each pair of successive x-values in the table. Enter 0.5 as the ΔTbl value and press [ENTER]. Use the arrow and [ENTER] keys to highlight "Auto" for both the dependent and independent variables so that the calculator will construct the table automatically.

Press [2nd] [Table] to display the completed table.

Use the arrow keys to scroll through the table entries. According to the table, after 4 seconds, the rocket will be 100 feet high.

X	Y_1
1	40
1.5	56.25
2	70
2.5	81.25
3	90
3.5	96.25
4	100

X = 4

Try This

1. Geothermal energy is generated whenever water comes in contact with hot underground rocks. The heat turns the water into steam that can be used to make electricity. The underground temperature of rocks varies with their depth below the surface. The temperature in degrees Celsius, y, is estimated by the equation $y = 35x + 20$, where x is depth in kilometers.

    a. Use a graphing calculator to create a table of values for $x = \{0, 5, 10, 15, 20, 25, 30, 35, ...\}$. **See the Teacher's Answer Key.**

    b. What would be the temperature of rocks that are 25 kilometers below the surface? **895°**

2. As a thunderstorm approaches, you see lightening as it occurs, but you hear the accompanying thunder a short time afterward. The distance in miles, y, that sound travels in x seconds is given by the equation $y = 0.21x$.

    a. Use a graphing calculator to create a table of values for $x = \{0, 1, 2, 3, 4, 5, 6, 7, ...\}$. **See the Teacher's Answer Key.**

    b. How far away is lightening when the thunder is heard 7 seconds after the light is seen? **1.47 miles**

Graphing Calculator Activity 3: Perimeter of Inscribed Polygons

The graphing calculator program below will determine the length of a side of a regular polygon inscribed in a circle and the perimeter of the polygon. In order to use the program, you must first enter the program into the calculator's memory. To access the program memory, use the following keystrokes.

Enter: [PRGM] [▶] [▶] [ENTER]

Example Find the length of one side and the perimeter of an octagon inscribed in a circle that has a radius of 5 centimeters.

Enter: [PRGM] 1 [ENTER]

Enter the number of sides and press [ENTER].

Then enter the radius and press [ENTER].

```
PROGRAM:SIDELENG
:Degree
:Disp "NUMBER OF
 SIDES"
:Input S
:Disp "RADIUS OF
 CIRCLE"
:Input R
:(180/S) →A
:(2*R*sin A) →L
:S*L →P
:Disp "SIDE LENGTH"
:Disp L
:Disp "PERIMETER"
:Disp P
:Stop
```

The program is written for use on a TI-82 graphing calculator. If you have a different type of programmable calculator, consult your User's Guide to adapt the program for use on your calculator.

```
RADIUS OF CIRCLE
?5
SIDE LENGTH
        3.826834324
PERIMETER
        30.61467459
```

The length of one side of the regular polygon is about 3.8 centimeters, and the perimeter is about 30.6 centimeters.

Try This

Use the program to find the length of one side and the perimeter of each polygon inscribed in a circle with the given radius or diameter.

1. Equilateral triangle; radius: 15 centimeters.   **25.98 cm; 77.94 cm**

2. Regular hexagon; radius: 2 inches.   **2 in.; 12 in.**

3. Square; diameter: 20 feet.   **14.14 ft; 56.57 ft**

4. Regular pentagon; radius: 300 meters.   **352.67 m; 1,763.36 m**

5. Regular octagon; diameter: 26.2 inches.   **10.03 in.; 80.21 in.**

Graphing Calculator Activity 4: Area of Sectors

The graphing calculator program below will determine the area of a sector of a circle that has been cut into equal sectors. In order to use the program, you must first enter the program into the calculator's memory. To access the program memory, use the following keystrokes.

Enter: [PRGM] [▶] [▶] [ENTER]

Example Find the area of a slice of a circular pizza that is cut into 8 pieces and has a diameter of 14 inches.

First set the range. The notation $[-15, 15]$ by $[-10, 10]$ means a viewing window in which the values along the x-axis go from -10 to 10 and the values along the y-axis go from -10 to 10.

The program is written for use on a TI-82 graphing calculator. If you have a different type of programmable calculator, consult your User's Guide to adapt the program for use on your calculator.

Enter: [WINDOW] [ENTER]

[(-)] 15 [ENTER] 15 [ENTER] 1 [ENTER]

[(-)] 10 [ENTER] 10 [ENTER] 1 [ENTER]

Now run the program.

Enter: [PRGM] 2 [ENTER]

```
PROGRAM:SECTOR
:FnOff
:PlotsOff
:ClrHome
:ClrDraw
:Disp "ENTER THE
 RADIUS"
:Input R
:Disp "NO. OF SE
 CTORS"
:Input P
:"√(R² − X²)"→Y₁
:"−√(R² − X²)"→Y₂
:"X"→Y₃
:"−X"→Y₄
:((π*R²)/P)→A
:Shade(Y₃,Y₁,1,0)
:Pause
:Disp "SECTOR AR
 EA IS"
:Disp A
:Stop
```

Enter the radius and press [ENTER]. Then enter the number of pieces and press [ENTER].

No matter how many pieces there are, you will see an illustration similar to the one shown at the right. Press [ENTER] and the calculator will display the area.

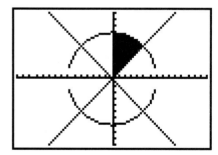

The area of one slice is about 30.6 centimeters.

● **Try This**

Use the program to find the area of one slice of a circular pizza with the given diameter.

1. 12 pieces; diameter: 12 inches
   about 9.4 in²

2. 6 pieces; diameter: 10 inches
   about 13.1 in²

3. 9 pieces; diameter: 14 inches
   about 17.1 in²

4. 4 pieces; diameter: 6 inches
   about 7.1 in²

Glossary

A

Apothem (p. 13)   The segment drawn perpendicular from the center of a regular polygon to any one of its sides

Example $\overline{AB}$ is an apothem of the pentagon below.

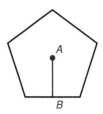

Arc 28)   Two points on a circle and all the points on the circle between them

Example A, B, and all the points between them form $\overparen{AB}$.

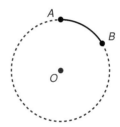

C

Central angle (p. 36)   An angle whose vertex is at the center of a circle

Example $\angle AOB$ is a central angle of circle O below.

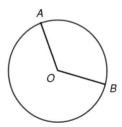

46   **Exploring Circles**

Chord (p. 36)   A segment joining any two points on the circle

Example $\overline{AB}$ is a chord of circle O below.

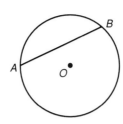

Circumference (p. 6)   The perimeter of a circle

Example   The circumference of a circle is equal to its diameter times π.

Concentric circles (p. 8, 27)   Circles with the same center but not necessarily the same radius.

Example   Three concentric circles are shown below

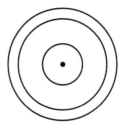

Coordinate geometry (p. 20)   The algebraic study of geometry through the use of a coordinate system

Example   In coordinate geometry, points are located using ordered pairs

---

D

Diameter (p. 6)   A straight line passing through the center of a circle from one side to the other

Example $\overline{AB}$ is the diameter of circle O below

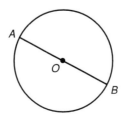

Glossary

I

Infinite product (p. 26)   The multiplication of an unlimited sequence of terms

Example $\frac{1}{1} \cdot \frac{2}{3} \cdot \frac{3}{4} \cdot \frac{4}{5}$ ...will result in an infinite product.

Infinite series (p. 26)   The summation of an unlimited sequence of terms.

Example   25 + 20 + 16 + 12.8 + ... is an infinite series.

Inscribed angle (p. 36)   An angle whose vertex is on a circle and whose sides are determined by two chords.

Example $\angle ABC$ is an inscribed angle of circle O below.

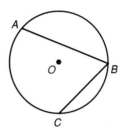

Intercepted arc (p. 36)   An arc of a circle that is cut off or bounded by two radii of that circle

Example $\widehat{XY}$ is the intercepted arc of $\angle XOY$ below.

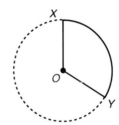

O

Ordered pair (p. 20)   The coordinates of a point in a plane

Example   Point A below is represented by the ordered pair (3, 2).

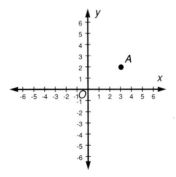

48   Exploring Circles

P

Perimeter (p. 3) The sum of the lengths of the sides of a geometric figure
Example If the length of the side of a square is 2 units, the perimeter is 8 units.

Polygon (p. 2) A simple closed plane figure.
Example Figure *ABCDE* below is a polygon.

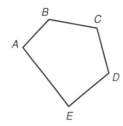

R

Regular polygon (p. 2) A polygon having all sides congruent and all angles congruent
Example Squares and equilateral triangles are regular polygons.

Radian (p. 28) The measure of an angle formed at the center of a circle of radius 1 unit that intercepts an arc whose length is 1 unit.
Example In the figure below, m∠*AOB* = 1 radian.

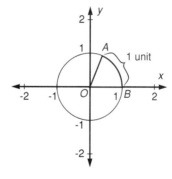

Glossary 49

S

Scale drawing (p. 2)   A picture that is drawn of a different size object with a given ratio between the dimensions of the representation of the object and that of the actual object

Example   The ratio in the scale drawing below is 1 unit = 1 foot.

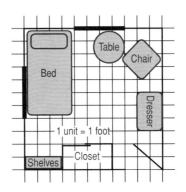

Sector (p. 30)   A region bounded by two radii and an arc of a circle

Example   *ACBO* is a sector of circle *O* below.

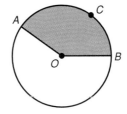

Slope (p. 6)   The ratio of the vertical change to the horizontal change of a line

Example   The slope of the line containing points (3, 1) and (5, 5) is

$$\frac{5-1}{5-3} = \frac{4}{2} = 2$$

X

X-coordinate (p. 20)   The first value of an ordered pair
    Example   Point $A(3, 2)$ has an x-coordinate of 3.

Y

Y-coordinate (p. 20)   The second value, or ordinate, of an ordered pair
    Example   Point $A(3, 2)$ has a y-coordinate of 2.

Index

A

Angles
    central, 36-37
    inscribed, 36-37
Apothem, 13
Archimedes, 18-20
Arcs, 28-30
    intercepted, 36
Area, 12-17
    circle, 14, 17, 21, 29-32
    polygon, 12-14
    rectangle, 14-17
    triangle, 13

C

Central angle, 36
Chords, 36
Circles
    arcs, 28-30, 36
    area, 14, 17, 21, 29-32
    circumference, 5-11
    concentric, 8-9, 27, 31-32
    diameter, 5-11
    sector, 30
Circumference, 5-11
Concentric circles, 8-9, 27, 31-32
Coordinate geometry, 20

D

Descartes, Rene, 20
Diameter, 5-11

F

Function, 11

G

Geometry, 2
Graphing calculator programs, 4, 24, 42-45

I

Infinite product, 26
Infinite series, 26
Inscribed angle, 36
Intercepted arc, 36

L

LOGO, 33-41

O

Ordered pair, 20

P

Perimeter
    polygon, 3, 18-19
    rectangle, 14-16
Pi (π), 9, 18-26
Polygons, 2
    apothem, 13
    area, 12-14
    creating, 34-36
    perimeter, 3-5, 18-19
    regular, 2

R

Radian, 28
Regular polygon, 2

S

Scale drawing, 2, 15
Sector, 30
Slope, 6
Spinner simulations, 22-25

X

X-coordinate, 20

Y

Y-coordinate, 20

Photo Credits

Cover, (bicycle sign) Bernd Kappelmeyer/FPG, (clocks) Dennis O'Clair/TSI, (pizza) Marvy!/The Stock Market; **1,** Life Images; **5,** Sue Norton; **11,** Bruce Ayres/TSI; **12,** Gabe Palmer/The Stock Market; **17,** Life Images; **21,** Aaron Haupt; **23,** Glencoe File; **27,** Peter Gridley/FPG; **32,** Joseph Devenney/Image Bank; **37,** Life Images; **41,** Finley-Holiday/FPG.